Essays in Biochemistry

volume 36 2000

Essays in Biochemistry

Molecular Trafficking

Edited by P. Bernstein

Portland Press

Essays in Biochemistry is published by Portland Press Ltd
on behalf of the Biochemical Society

Portland Press
59 Portland Place
London W1B 1QW, U.K.
Fax: 020 7323 1136; e-mail: editorial@portlandpress.com
www.portlandpress.com

In North America orders should be sent to Princeton University Press,
41 William Street, Princeton, NJ 08540, U.S.A.

All profits made from the sale of this publication are returned to the Biochemical Society for the promotion of the molecular life sciences.

British Library Cataloguing-in-Publication Data
A catalogue record for this book is available from the British Library

ISBN 1 85578 131 X
ISSN 0071 1365

Typeset by Portland Press Ltd
Printed in Great Britain by Information Press Ltd, Eynsham, U.K.

Contents

3 Glycosylation and protein transport

Peter Scheiffele and Joachim Füllekrug

4 Vesicular transport

Francis Barr

5 Functions and origins of the chloroplast protein-import machinery

Danny J. Schnell

6 Mechanisms of mitochondrial protein import

Donna M. Gordon, Andrew Dancis and Debkumar Pain

7 Pore relations: nuclear pore complexes and nucleocytoplasmic exchange

Michael P. Rout and John D. Aitchison

8 Nuclear transport: never-ending cycles of signals and receptors

Dianne M. Barry and Susan R. Wente

9 The control of gene expression by regulated nuclear transport

Eric D. Schwoebel and Mary Shannon Moore

10 RNA export: insights from viral models

Matthew E. Harris and Thomas J. Hope

Preface

All cells carry out a programme of renewal, continually degrading and synthesizing their components. As each new molecule is synthesized it must find its way to a specific location within the cell. Often the targets are membrane-bound compartments called organelles. The lipid membrane surrounding each organelle serves to keep some macromolecules in — for example, proteins and nucleic acids — while excluding others. This level of cellular organization allows a defined set of biochemical reactions to be carried out within each specific compartment. The synthesis of RNA from its DNA template, for example, takes place within the nucleus of eukaryotic cells, whereas, in plant cells, the biochemical reactions of photosynthesis are carried out within chloroplasts. Protein synthesis takes place within the cytoplasm, whereas RNA is synthesized, for the most part, within the nucleus. Proteins and RNA often carry out their functions within cellular compartments different from those in which they were synthesized and must, therefore, traverse lipid membranes that are normally impermeable. As a result, the cell must be able to transport these macromolecules across lipid membranes to the site where they ultimately function.

The topic of this volume of *Essays in Biochemistry* is molecular trafficking. The 10 review articles that follow discuss the mechanisms and machinery that facilitate the trafficking of proteins and nucleic acids across different organelles within a variety of cell types. The timeliness of this volume is exemplified by the fact that the 1999 Nobel Prize in Physiology or Medicine was awarded to Dr Günter Blobel for the discovery that "proteins have intrinsic signals that govern their transport and localization in the cell". Dr Blobel's Nobel Prize-winning research demonstrated that proteins contain specific intrinsic signals — address tags — which govern their transport to specific intracellular organelles. As illustrated in the following essays, this simple signal hypothesis has been used by cells to solve a diverse set of problems from transport of proteins into the endoplasmic reticulum, chloroplasts or mitochondria, to shuttling of proteins and nucleic acids in and out of the nucleus.

The basic theme — that a specific signal contained within each macromolecule is recognized by a cellular machine to transport that molecule across a pore within the lipid membrane — is first described in Chapter 1 and reiterated in subsequent chapters. There is substantial diversity to this molecular theme. Not only do the signals differ, as would be expected for address tags, but, in addition, the transporters themselves share only a superficial similarity. Membrane pores seem to be a ubiquitous feature, but their characteristics differ depending on the organelle in which they are present. This is due, in part, to the structural differences between each of the organelles, i.e. the complex lipid bilayer of the nucleus versus the cisternal organization of the mitochondrion. The differences in pore structures may also further regulate the types of molecule that traverse the different membranes and the directionality of the transport. As will be seen in Chapters 7, 8 and 10, the complexity of the nuclear pore and its associated transport machinery may be due to the necessity to transport chemically diverse molecules — proteins and nucleic acids — in a bi-directional manner (in and out of the nucleus). Chloroplasts and mitochondria (Chapters 5 and 6) generally only transport proteins in a single direction; however, these structurally complex organelles have had to solve the problem of targeting proteins across one or several membranes of the cisternae. Although their pore structure appears less complex than the nuclear pore, a unique set of mechanisms and machinery allows for the regulated targeting required by these organelles. The address tags themselves can be modified. In Chapter 3, glycosylation, a post-translational event that specifically allows a subset of proteins to be targeted to different intracellular or extracellular locations, is described. Similarities between molecular targeting and secretion are examined in Chapter 4 in the context of bulk flow. Although pore structures are not involved in secretory vesicle trafficking, some of the principles used to target proteins to specific organelles also apply to the transport of vesicles. Chapters 2 and 9 discuss how molecular trafficking shapes specific biological events, immunological surveillance and gene regulation, respectively.

As each cell and organelle issues its own diverse set of problems in targeting newly synthesized macromolecules, it will be fascinating to learn how the common themes discussed in this volume continue to develop and diverge.

Philip Bernstein (New York)
August 2000

Authors

Suzanna Meacock studied Biochemistry at the University of Manchester and stayed in Manchester to carry out her Ph.D. in Stephen High's laboratory. Suzanna's work focused on the biogenesis of polytopic integral membrane proteins and she was awarded her Ph.D. in January 2000. Suzanna is presently training to become a patent attorney. **Julie Greenfield** carried out her Ph.D. studies at the Institute of Arable Crops Research, Long Ashton, Bristol, before joining Stephen High's group as a postdoctoral research associate. Julie's postdoctoral research focused on the structure/function of the endoplasmic reticulum and in particular on the subcellular localization of ER translocon components. Julie is now working in biomedical publishing. **Stephen High** has a long-standing interest in membrane-protein biogenesis at the endoplasmic reticulum. This originated from his Ph.D. studies of red-cell membrane proteins and continued during his time as an EMBO fellow with Bernhard Dobberstein. For the past decade Stephen has been studying various aspects of membrane-protein biosynthesis with his own research group in Manchester. His current interests include membrane insertion, chaperone-mediated protein folding and the relationship between the misfolding of membrane proteins and specific diseases. He is currently a Professor of Biochemistry at the University of Manchester.

Christopher Nicchitta received his B.Sc. in Biology at the College of William and Mary in 1981 and his Ph.D. in Biochemistry/Biophysics from the University of Pennsylvania in 1987. He then moved to Rockefeller University, New York, where he was a postdoctoral fellow in Dr Günter Blobel's laboratory from 1988 to 1993. Following his postdoctoral studies, he moved to Duke University as an Assistant Professor of Cell Biology and was promoted to Associate Professor in 2000. **Robyn Reed** received her B.Sc. in Biology at Wake Forest University in 1996. She is currently a fourth-year M.D./Ph.D. student at Duke University, working towards a Ph.D. in Cell Biology.

Peter Scheiffele studied Biochemistry at the Freie Universität Berlin. During his Ph.D. in the laboratory of Kai Simons at EMBL, Heidelberg, he analysed mechanisms of apical transport in polarized epithelial cells. He is currently a postdoctoral worker at the University of California San Francisco in the field of cellular neurobiology. **Joachim Füllekrug** obtained a degree in Chemistry from the University of Göttingen. After 1 year of Molecular Biology at the University of Kent at Canterbury he decided to become a molecular cell biologist. Always interested in secretion, he has worked his way up the pathway, starting with resident ER proteins during his Ph.D., then focusing on ER-Golgi cycling proteins during his postdoctoral studies at EMBL, and is now investigating sorting at the *trans*-Golgi network.

Francis Barr did his Ph.D. with Professor Wieland Huttner in the Cell Biology programme of the EMBL in Heidelberg, looking at the formation of secretory granules from the *trans*-Golgi network in neuroendocrine cells. While working as a postdoc in the laboratory of Graham Warren at the ICRF in London he started to look for proteins involved in establishing the stacked structure of the Golgi apparatus, which lead to the discovery of the GRASP proteins. He then became an independent researcher at the University of Glasgow, and has recently moved to the Max Planck Institute for Biochemistry in Munich to take up a position as a group leader.

Danny J. Schnell is Associate Professor of Cell Biology in the Department of Biological Sciences at Rutgers University. He received his Ph.D. at the University of California, Davis, and trained as a postdoctoral associate with Dr G. Blobel at the Rockefeller University, New York. His research interest is plant cell biology with specific focus on plastid biogenesis.

Donna Gordon obtained her Ph.D. in Cell and Molecular Biology in 1998 from the University of Pennsylvania School of Medicine. She is currently a postdoctoral fellow in the laboratory of Dr Debkumar Pain and is studying the role of GTP in mitochondrial protein import. **Andrew Dancis** obtained his M.D. degree from New York University in 1978. This was followed by residency in Internal Medicine and fellowship in Hematology. In 1986, he joined the laboratory of Dr Richard Klausner at the National Institutes of Health. In 1996, he joined the Department of Medicine at the University of Pennsylvania as Assistant Professor. His current research interest includes the genetics of iron metabolism in yeast. **Debkumar Pain** is Assistant Professor of Physiology at the University of Pennsylvania School of Medicine. He obtained his Ph.D. in Biochemistry from the University of Calcutta, India. He worked with Dr Günter Blobel at the Rockefeller University, New York, on various aspects of protein translocation into organelles. His current research interests include mitochondrial biogenesis and functions.

Michael Rout is a Rita Allen Foundation Scholar, a recipient of an Irma T. Hirschl Career Scientist Award, and Assistant Professor and Head of the Laboratory of Cellular and Structural Biology at the Rockefeller University in New York. He obtained his Ph.D. at the MRC Laboratory of Molecular Biology and the University of Cambridge in 1989, and then joined Günter Blobel's laboratory at Rockefeller University for postdoctoral studies. During his Ph.D. he isolated the spindle organizer from yeast, and building on this expertise isolated the yeast nuclear pore complex in Blobel's laboratory. **John Aitchison** is an Assistant Professor and MRC and Heritage Scholar at the University of Alberta in Canada. He obtained his Ph.D. at McMaster University in Hamilton, Canada, in 1992 and then joined Günter Blobel's laboratory at Rockefeller University. During his Ph.D. studies under the direction of Dr Richard Rachubinski, he developed a molecular assay system to study peroxisomal biogenesis in the yeast. Together in Blobel's laboratory, Drs Aitchison and Rout collaborated on several projects to identify and characterize components of the yeast nuclear pore complex and many novel transport factors that mediate protein import into the nucleus. In their own laboratories, Dr Aitchison is concentrating on the soluble factors while Dr Rout continues to study the structure of the nuclear pore complex.

Dianne Barry received a Ph.D. in Neurobiology from Washington University School of Medicine in 1997 after having completed her thesis work on the molecular correlates of the mammalian cardiac current, I_{to}. She completed postdoctoral training in the laboratory of Dr Susan Wente, where she examined the role of the nuclear pore complex-associated protein, hGle1p, in the vertebrate mRNA-export pathway. She is currently employed by the Bayer Chemical Company. **Susan Wente** is an Associate Professor of Cell Biology and Physiology at Washington University School of Medicine. She received her Ph.D. in Biochemistry from the University of California, Berkeley, in 1988 with Dr Howard Schachman. After a 1-year fellowship with Dr Ora Rosen at the Memorial Sloan Kettering Cancer Center, she began her studies of nuclear transport with Dr Günter Blobel at The Rockefeller University, New York. In 1993, Dr Wente was named an Assistant Professor at the Washington University School of Medicine, and was promoted to Associate Professor in 1998. Recent research efforts have focused on using yeast and vertebrate model systems to understand the mechanism of nuclear transport and the pathway of nuclear pore complex assembly.

Mary Shannon Moore is an Assistant Professor in the Department of Molecular and Cellular Biology at Baylor College of Medicine in Houston. She received her Ph.D. from the University of Texas Southwestern Medical Center in Dallas and did a postdoctoral fellowship at Rockefeller University, New York, prior to joining the Baylor faculty. **Eric D. Schwoebel** received his Ph.D. at Baylor College of Medicine, worked at the Institute for Primate Research in Nairobi, Kenya, as a senior research scientist, and is currently a postdoctoral fellow in Dr Moore's laboratory.

Matthew E. Harris received his Ph.D. from the Biology Department at the University of California, San Diego, in 1999. He received his B.Sc. in Biology from Harvey Mudd College in 1993. Presently, he is a commissioned officer in the United States Army serving at the Walter Reed Army Institute of Research. **Thomas J. Hope** is an Assistant Professor at the Salk Institute for Biological Studies in the Infectious Disease Laboratory. Dr Hope received his Ph.D. from the University of California, Berkeley, in 1988. He was a postdoctoral fellow at the University of California, San Francisco, from 1988 to 1992, where he started his studies of the post-transcriptional regulation of viral gene expression. Presently, his research is directed towards understanding RNA processing and export using viral models on intronless messages.

Abbreviations

cNLS	classical nuclear localization sequence
CTE	constitutive transport element
$\Delta\Psi$	membrane potential
ER	endoplasmic reticulum
ERGIC-53	ER-Golgi intermediate-compartment protein of 53 kDa
GIP	general insertion pore
GR	glucocorticoid receptor
HBV	hepatitis B virus
HIV	human immunodeficiency virus
hnRNP	heterogenous nuclear ribonucleoprotein
HSV	herpes simplex virus
IM	inner membrane
IMS	intermembrane space
LHCP	light-harvesting chlorophyll a/b-binding protein
LMB	leptomycin B
LR-NES	leucine-rich nuclear export sequence
LTD	thylakoid luminal targeting domain
MDCK	Madin–Darby canine kidney
MHC	major histocompatability complex
MPMV	Mason–Pfizer monkey virus
MPP	matrix-processing peptidase
MSF	mitochondrial-import-stimulating factor
mt-Hsp70	mitochondrial Hsp70
NE	nuclear envelope
NES	nuclear export sequence
NLS	nuclear localization sequence
NPC	nuclear pore complex
NSF	*N*-ethylmaleimide-sensitive fusion protein
NUP	nucleoporin
OM	outer membrane
PDI	protein disulphide isomerase
PPE	pre-mRNA processing enhancer
RanBP1	Ran-binding protein 1
RanGAP	Ran-GTPase-activating protein
RNP	ribonucleoprotein
RRE	Rev-responsive element
SNAP	soluble NSF attachment protein
SNARE	soluble NSF attachment protein receptor

snRNA	small nuclear RNA
snRNP	small nuclear ribonucleoprotein
SREBP	sterol regulatory-element-binding protein
SRP	signal-recognition particle
SRP54	54 kDa polypeptide of SRP
STD	stromal-targeting domain
TAP	transporter associated with antigen presentation
TK	thymidine kinase
TRAM protein	translocating-chain-associating membrane protein
VIP36	vesicular integral membrane protein of 36 kDa

1

Protein targeting and translocation at the endoplasmic reticulum membrane — through the eye of a needle?

Suzanna L. Meacock, Julie J.A. Greenfield and Stephen High[1]

School of Biological Sciences, University of Manchester, 2.205 Stopford Building, Oxford Road, Manchester M13 9PT, U.K.

Introduction

A distinguishing feature of eukaryotic cells is the presence of membrane-bound organelles within the cytoplasm. These organelles have distinct functions that are reflected by the presence of specific proteins within their membrane(s), and a unique luminal composition. Newly made proteins are targeted from their site of synthesis, the cytosol, to the appropriate organelle by virtue of specific signal sequences. These act as intracellular 'address labels' and ensure that a precursor protein is delivered to the correct location. These signal sequences usually take the form of amino acid motifs within the precursor protein, and distinct signals for targeting proteins to the nucleus, mitochondrion, peroxisome and endoplasmic reticulum (ER) have been identified. The ER plays a key role within the cell since it is the entry point into the 'secretory pathway', providing access to all of the compartments of that pathway and the cell surface. This chapter will focus on protein targeting

[1]*To whom correspondence should be addressed (e-mail: SHigh@fs1.scg.man.ac.uk).*

and translocation at the ER of mammalian cells, drawing on studies of other organisms where these are better understood. When a live mammalian cell is viewed under a microscope, one sees that the ER is closely associated with the nucleus and forms a reticular network that stretches to the periphery of the cell (see Figure 1).

ER signal sequences

For proteins destined to enter the ER, the signal sequence, or targeting motif, is particularly well characterized and defined as a continuous stretch of 6–20 hydrophobic amino acid residues usually located towards the N-terminus of the protein, and often flanked by one or more basic residues [1]. ER signal sequences may either be cleaved following protein targeting, or remain an integral part of the mature protein. Cleavable signal sequences are removed during translocation across the ER membrane by the action of the signal peptidase complex located on the luminal side of the membrane. Many secretory proteins have cleavable signal sequences (Figure 2).

Not all proteins possessing cleavable signal sequences are destined for the ER lumen, and many are integrated into the ER membrane. The membrane integration of such proteins is achieved by the presence of a second stretch of hydrophobic amino acids, C-terminal to the cleavable signal sequence. This region is called a 'stop-transfer' sequence since it acts to stop the translocation, or transfer, of the protein across the membrane, and functions as the transmembrane domain of the mature protein (Figure 2). Many other integral mem-

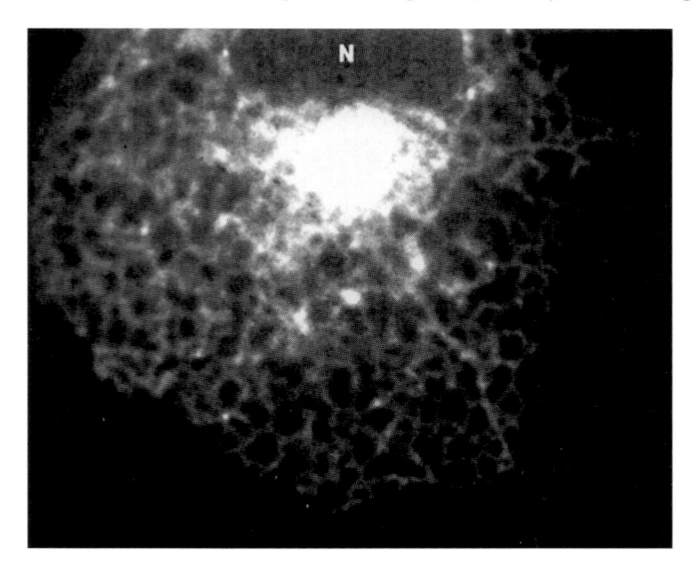

Figure 1. The ER of a living cell
The ER of a living COS-1 mammalian cell visualized via a green fluorescent protein-tagged version of the Sec61α protein (see [30]). A region adjacent to the nucleus (N) is intensely labelled and the reticular network characteristic of the ER can also be seen.

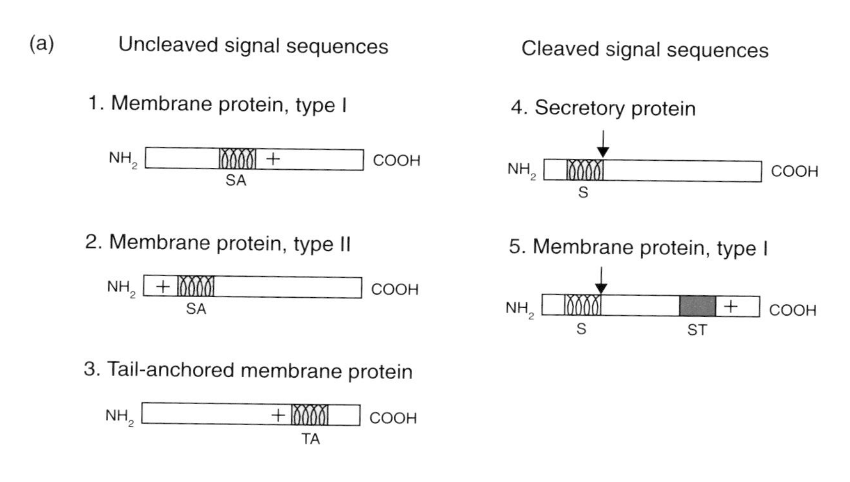

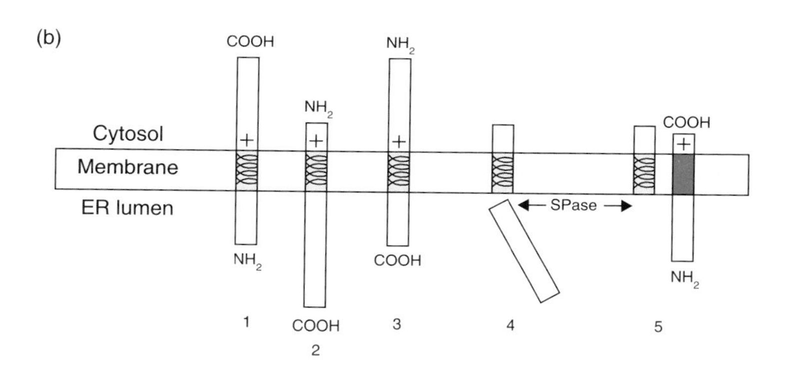

Figure 2. ER targeting signals and their resulting transmembrane topologies
Hydrophobic ER signal sequences are depicted by a coiled motif. (a) S denotes a cleavable signal sequence, SA a signal-anchor sequence, ST a stop-transfer sequence and TA a tail-anchor sequence. Charged regions flanking these hydrophobic sequences can influence the final orientation of transmembrane proteins [1] and are represented by a plus sign. (b) A type-I membrane protein retains its C-terminus in the cytosol while a type-II membrane protein retains its N-terminus in the cytosol [1]. Where appropriate, the cleavage of a signal sequence by signal peptidase (SPase) is indicated by an arrow.

brane proteins contain uncleaved signal sequences (Figure 2). In this case, a single 'signal-anchor' sequence serves two functions, i.e. targeting the protein to the ER and acting as the transmembrane anchor [1].

A special class of integral membrane protein has an uncleaved signal sequence located near the extreme C-terminus of the polypeptide (Figure 2). These have been denoted 'tail-anchored' proteins and their biosynthesis is quite distinct from all the other ER-targeted proteins described here [2]. In particular, these proteins are inserted post-translationally via a novel mechanism that is poorly characterized and which falls outside the scope of this review.

Signal recognition particle

The synthesis of proteins destined for the ER begins on cytosolic ribosomes. As the ER signal sequence is usually located towards the N-terminus of the protein, the signal exits the ribosome at an early stage of synthesis. This exposed signal sequence is now available for 'sampling' by the cytosolic signal-recognition particle (SRP; Figure 3) [3]. Mammalian SRP is a ribonucleo-protein complex comprising six different polypeptides assembled on a 7 S RNA molecule [4]. The most important protein subunit of SRP is the 54 kDa polypeptide (SRP54), and it is this subunit that recognizes and binds to the hydrophobic ER targeting signals. The structure of the SRP54 subunit is crucial to its role, and three domains have been identified on the basis of several criteria [4]. These are the N-terminal N domain, the central G domain, which can bind and hydrolyse GTP, and the C-terminal methionine-rich M domain that binds to both the hydrophobic ER targeting signals and the 7 S RNA component [4]. The binding of the ER signal sequence by SRP54 allows other subunits of the SRP to mediate a slowing down in the rate of protein synthesis [4]. This reduction in translation rate by the SRP prevents large amounts of the new protein becoming exposed to the cytosol and perhaps developing a tightly folded tertiary structure that would interfere with the subsequent translocation of the nascent polypeptide across the ER membrane.

Homologues of SRP54 have been identified in a wide range of organisms including mycoplasma, bacteria and yeast, and in each case the homologue is

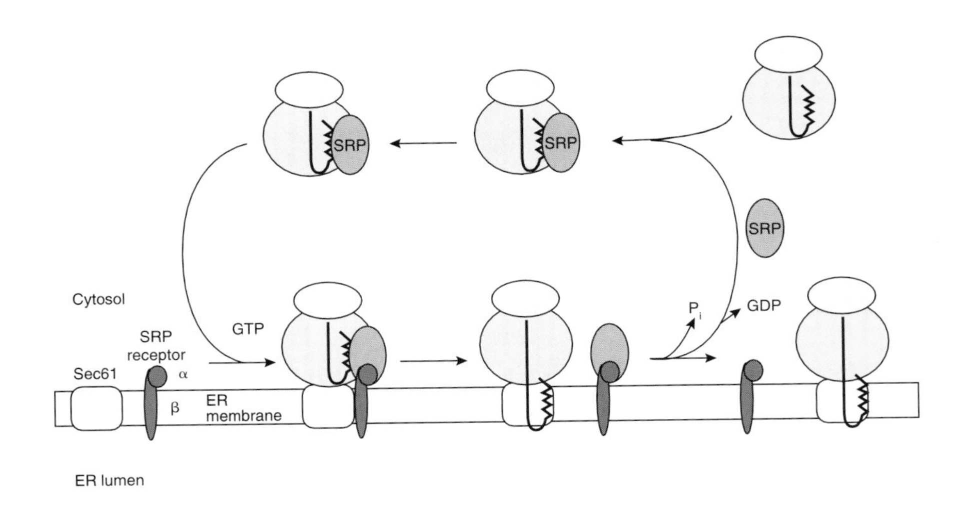

Figure 3. SRP-dependent targeting to the ER membrane
SRP binds to an ER targeting signal (represented by a zig-zag line) as it emerges from the ribosome and delivers the ribosome–nascent chain–SRP complex to the ER membrane via an interaction with the SRP-receptor complex. The signal sequence is released from SRP and interacts with the Sec61 complex, the core component of the ER translocation site. This SRP-dependent targeting cycle is regulated by GTP binding and hydrolysis (see text).

involved in protein targeting. The crystal structure of the *Thermus aquaticus* SRP54 homologue subdomains have been solved and are thought to be representative of all SRP54 homologues. The G domain is clearly related to other GTP-binding proteins while the N domain may sense or control nucleotide binding by the G domain [5]. The most prominent feature of the *T. aquaticus* SRP54 homologue M domain is a deep grove lined by the side chains of hydrophobic amino acids: this is almost certainly the signal-sequence-binding pocket of the M domain [6].

SRP-dependent targeting to the ER membrane

Once the SRP has bound to the ER targeting signal present on a short, incomplete, polypeptide chain emerging from the ribosome, this ribosome–nascent chain–SRP complex is specifically targeted to the ER membrane by virtue of an interaction between SRP and the SRP receptor (Figure 3). The binding of the SRP to a ribosome–nascent-chain complex, followed by the specific interaction of this complex with the SRP receptor, provides an efficient targeting mechanism for the presentation of nascent polypeptides at the ER translocation site (Figure 3). It appears that this SRP-dependent targeting route is not inhibited by the binding of non-translating ribosomes to the ER translocation site [7,8], although there is by no means agreement on this point [9].

The SRP receptor is a heterotrimer that is restricted to the ER membrane by virtue of its β subunit, which has a transmembrane domain. The α subunit of the SRP receptor associates peripherally with the cytosolic face of the ER membrane via its tight interaction with the SRP receptor's β subunit [4]. In fact, the transmembrane domain of the β subunit can be deleted without a loss of function. This suggests that the SRP receptor need only be transiently associated with the ER membrane in order to carry out its function; presumably this can take place via interactions with components other than the SRP receptor β subunit [10]. Like the SRP54 subunit, both subunits of the SRP-receptor complex can bind GTP; indeed the primary and tertiary structures of the N and G domains of prokaryotic SRP54 and SRP-receptor α-subunit homologues are closely related [11]. The binding of GTP to both the SRP receptor's α subunit [12] and β subunit [11] is necessary for function (Figure 3).

Currently, the most widely accepted model of SRP-dependent targeting to the ER membrane is that the GTP-binding site of the SRP54 protein is empty when it binds to an ER targeting signal [4,12], albeit that this view is not universal [13]. Upon arrival at the ER membrane, SRP54 will interact with the SRP-receptor α subunit, which also has an empty GTP-binding site at this stage [12]. The proposal that the GTP-binding sites of SRP54 and the SRP-receptor α subunit are both empty at the start of the 'SRP cycle' (Figure 3) is strongly supported by the crystal structures of their prokaryotic homologues. The G domains of both proteins appear to have a stable tertiary structure even in the absence of any bound nucleotide [5,11]. Following the interaction of SRP54 with the SRP-receptor α subunit, the co-operative binding of GTP to both of these proteins stabilizes the SRP–SRP-receptor complex and initiates the transfer of the nascent chain's signal sequence from SRP54 to the proteins of the ER translocation site [12]. The subsequent hydrolysis of GTP by both SRP54 and the SRP-receptor α subunit leads to the dissociation of the complex and regenerates both components for new rounds of ER targeting (Figure 3). Thus the components of the SRP-dependent targeting cycle use GTP binding and hydrolysis to (i) regulate the delivery of nascent precursor proteins to the ER membrane, (ii) optimize their presentation to the ER translocation complex and (iii) recycle the targeting components.

SRP-independent ER targeting

It is clear that some proteins with apparently 'classical' N-terminal ER targeting signals can be targeted to the ER membrane independently of SRP binding. In higher eukaryotes such proteins are generally short secretory proteins of around 70 amino acids where the synthesis of the protein is completed just as the signal sequence is emerging from the ribosome and therefore available for SRP binding. In eukaryotes SRP seems to function solely in mediating the targeting of nascent, ribosome-bound, polypeptides and cannot therefore assist the targeting of these short proteins [14]. SRP-independent targeting to the ER membrane requires ATP and cytosolic chaperones including members of the Hsp70 family [15].

The SRP-independent pathway seems to be used extensively in the yeast *Saccharomyces cerevisiae*, where the genetic elimination of the SRP-dependent targeting route does not completely prevent the targeting of proteins to the ER membrane. Yeast can survive without a functional SRP or SRP receptor, suggesting that there is at least one other pathway that allows proteins to be directed to the ER. Whereas yeast can survive in the absence of SRP-dependent protein targeting, the growth of the cells is severely diminished. Nevertheless, one must conclude that alternative targeting routes can deliver enough protein to the ER to enable cell survival. Studies *in vivo* in yeast indicate that subtle features of the ER targeting signal define which targeting route a particular precursor uses for delivery to the ER membrane [16].

Membrane insertion

After SRP-bound ribosome nascent chains are released from the SRP, the nascent chain interacts with the ER translocation machinery. During insertion into the ER, the nascent chain has two possible destinations: it can be completely translocated across the ER membrane and enter the lumen as a secretory protein; alternatively, only part of the protein may be translocated across the membrane leaving other regions exposed to the cytosol. In the latter case the protein is left spanning the ER membrane. Essentially all membrane proteins with a cleavable ER signal sequence assume a type-I orientation in the membrane, where the N-terminus of the protein is translocated into the ER lumen while the C-terminus remains in the cytosol (Figure 2).

The final topology of membrane proteins with an uncleaved, signal-anchor, sequence is generally governed by the 'positive inside' rule. This states that basic residues adjacent to a signal-anchor domain will remain on the cytosolic side of the ER membrane (see [1] and Figure 2). Hence, net positive charge N-terminal of a transmembrane domain results in the C-terminus being translocated (type-II orientation, see Figure 2), whereas net positive charge C-terminal of a transmembrane domain results in the N-terminus being translocated (type-I orientation, see Figure 2). In fact, other factors such as the length and hydrophobicity of the transmembrane domain and the presence of extensive secondary structure can also influence the final transmembrane orientation of a membrane protein with a signal-anchor sequence [17]. Regardless of transmembrane orientation, or the number of times a protein spans the membrane, all membrane proteins except those with a tail anchor (see above) appear to be integrated by the classical ER translocation complex [1,18].

The translocation site of the ER membrane

A number of studies using assays *in vitro* and yeast genetics have been carried out by several laboratories and have led to the identification of the components of the ER translocation site [3,18–20]. It is worth emphasizing that this ER translocation site or 'translocon' is responsible for both the translocation of secretory proteins and the integration of membrane proteins, dependent only upon the specific signal sequences a particular precursor protein possesses.

The structure of the ER translocon has been established at low resolution. These studies show that the Sec61 complex (see below) forms an oligomeric ring structure in the ER membrane with a central pore of at least 20 Å ([21] and references therein). This oligomer constitutes an aqueous pore that spans the entire ER membrane, and which has an alternative conformation where the pore diameter increases to 60 Å [21]. Furthermore, the central pore of the ER translocon appears to align precisely with the site where a newly synthesized polypeptide emerges from the ribosome [22]. The first evidence that protein translocation across the ER membrane took place via a water-filled pore came from electrophysiological studies carried out in the laboratory of the Nobel laureate Günter Blobel [23]. The subsequent structural studies also included a contribution from Blobel and colleagues [22] and allowed us to actually see the aqueous pore [21].

Components of the translocation site

The water-filled protein-lined channel that crosses the ER membrane is formed by oligomers of the Sec61 complex [21]. The mammalian Sec61 complex is a heterotrimer composed of the Sec61α, Sec61β and Sec61γ subunits and the complex is essential for protein translocation and integration at the ER membrane (Figure 4) [19]. The mammalian Sec61α subunit appears to form the major component of the water-filled transmembrane channel through which proteins are transported across and integrated into the ER membrane (see Figure 5 and [24]). The mammalian Sec61β subunit has recently been shown to facilitate co-translational translocation at the ER membrane and may also recruit the signal peptidase complex into a transient association with the ER translocation site [25]. The exact role of the mammalian Sec61γ subunit remains to be elucidated although studies of *S. cerevisiae* indicate that this component has a crucial role in protein translocation [3]. Both Sec61α and Sec61β extend up into the docked ribosome, and are therefore adjacent to the translocating polypeptide as it passes through a continuous channel that starts inside the ribosome and extends through the ER membrane to its luminal face [3,19,22].

Despite the aqueous nature of the ER translocation channel (Figure 5), when hydrophobic targeting signals enter this channel they appear to have direct lateral access to the phospholipids of the membrane bilayer (Figure 5) [19,20]. To date, the structure of the ER translocon provides no clue as to how this access to the lipid phase is achieved [21,22]. Nevertheless, such access would greatly facilitate the transfer of hydrophobic transmembrane regions out of the translocation site and into the bilayer (see Figure 5).

In addition to the subunits of the Sec61 trimer, a second component, the translocating-chain-associating membrane (TRAM) protein, has been shown to be adjacent to specific regions of nascent polypeptides during their insertion into the ER translocation site [19,21]. In contrast to the Sec61 complex, the TRAM protein is restricted to higher eukaryotes. Although TRAM is not essential for the membrane translocation and integration of all proteins, it is required by many precursors and stimulates this process for many others [3,19]. Although the exact function of the TRAM protein has yet to be estab-

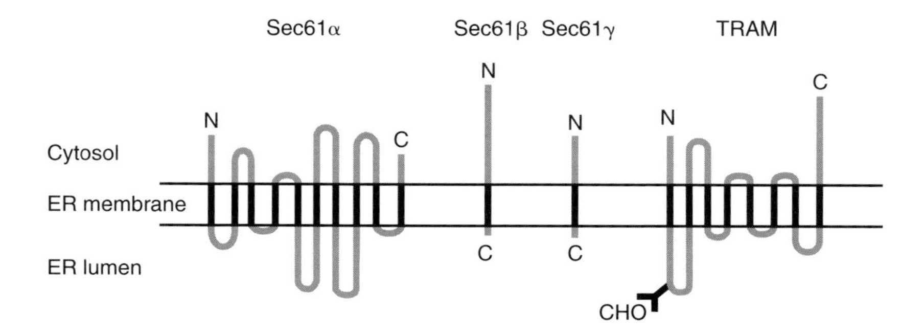

Figure 4. Core components of the mammalian ER translocon
The membrane topology of the subunits of the Sec61 complex and the translocating-chain-associating membrane (TRAM) protein are indicated. The TRAM protein is N-glycosylated (CHO).

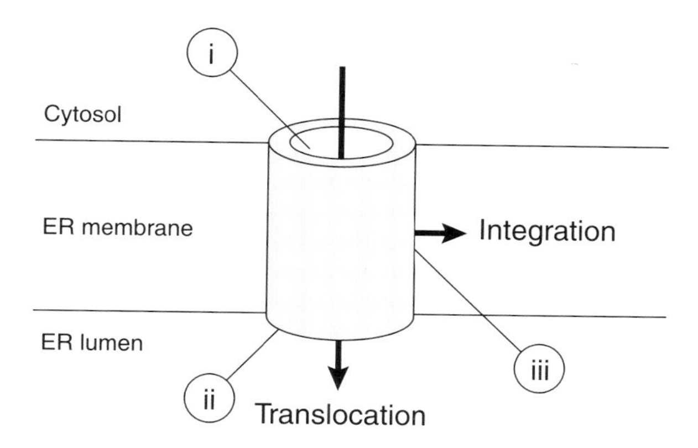

Figure 5. Locations where the water-filled ER translocon is 'gated'
The cylindrical pore represents an ER translocation site constructed from oligomers of the Sec61 complex. The directions of translocation and integration are indicated by arrows. Locations at which access to and from the water-filled translocon is regulated (gated) are as follows: (i) on the cytoplasmic face of the ER membrane by ribosome binding [27] and the translocating-chain-associating membrane (TRAM) protein [26]; (ii) on the luminal face of the ER membrane by BiP [28] and (iii) within the plane of the membrane [19,20,26].

lished, it is clear that the interaction of the TRAM protein with a nascent precursor is dependent upon the properties of its signal sequence [21]. The most recent suggestion for a function of the TRAM protein is that it can regulate the access of a nascent polypeptide within the ER translocation channel to the cytosol (see Figure 5 and [26]).

Gating of the ER translocation channel

During the early stages of membrane translocation, the ribosome creates a tight seal with the ER translocon and thereby closes the cytosolic face of the channel and promotes vectorial transport of the nascent chain through the translocon [27]. The luminal end of the ER translocon can also be closed in a regulated manner and this function is carried out by the ER chaperone BiP [28]. BiP can block the luminal end of both active (translocating) and inactive (empty) translocons, it therefore plays a key role in maintaining the permeability barrier of the ER membrane and in preserving the gradients established across it, for example that of Ca^{2+} [28]. The opening and closing of the seals at the two ends of the ER translocon is tightly regulated in response to signals such as nascent chain length and the presence of a transmembrane region in the ribosome [27,28]. This system ensures that both ends of the translocon are never open at the same time and hence maintains the various chemical gradients established across the ER membrane.

It seems clear that the lateral exit of a hydrophobic transmembrane domain, or cleavable signal sequence, from the ER translocon into the centre of the phospholipid bilayer (Figure 5) must also be controlled or regulated in some fashion [19,20]. Whereas it has been suggested that Sec61β [19] or the TRAM protein [28] may have some role in this process, exactly how such lateral exit is controlled is far from clear.

Inside out?

The ER is a major site of protein synthesis and it can recognize misfolded proteins and unassembled protein subunits by a process generally known as 'quality control'. It has recently become apparent that these misfolded proteins are returned from the ER to the cytosol where they are degraded by the proteasome-mediated pathway [29]. Clearly this first requires the proteins to be moved in a retrograde fashion out of the ER and into the cytosol. Several studies indicate that the Sec61 complex is at least partly responsible for this retrograde transport or 'dislocation' of misfolded proteins from the ER to the cytosol [29]. Quite how the Sec61 complex can co-ordinate its role in transporting proteins in two directions is unknown. One possibility is that forward translocation occurs at the ER, while retrograde translocation takes place at a distinct spatial location within the cell, namely the ER–Golgi intermediate compartment [30].

Accessory components at the ER translocation site

While the complexity of the ER translocon should now be apparent, we have so far restricted our description to the key players in the process of translocation. In fact it has long been known that many of the proteins entering the ER undergo alterations and modifications. Furthermore, most of these changes can occur while the protein is being inserted into or translocated across the membrane. This means that the ER components which carry out these changes must be closely associated with the translocon, albeit that they are not required for the actual translocation process [31] and can therefore be viewed as accessory components. Accessory components like the signal peptidase and oligosaccharyltransferase are multisubunit protein complexes that do not co-purify with the core components of the ER translocon [3] (Figure 6). It is not clear how all of these components can gain access to newly translocating chains to carry out their modification functions [31]. It also remains to be established whether each ER translocon is associated with a full set of accessory components, or whether a single accessory component can service several active ER translocons, associating transiently with each in turn.

Outlook

The future will bring great refinement to our understanding of protein synthesis at the ER. We can look forward to knowing how GTP-binding proteins co-ordinate the SRP-dependent targeting process. A high-resolution structure of the ER translocon core must surely come and will provide the

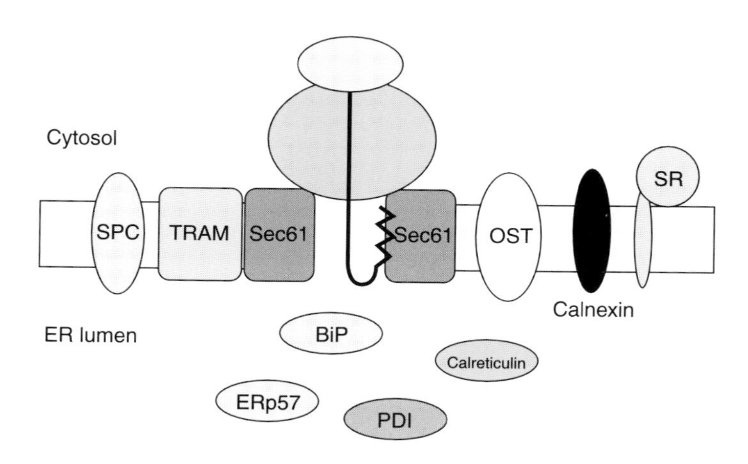

Figure 6. Accessory components of the ER insertion site
In addition to the core components of the ER translocation site, i.e. the Sec61 complex and the TRAM protein, the SRP-receptor complex (SR) and a number of accessory components must be closely associated with the site during translocation. These accessory components include the signal peptidase complex (SPC), the oligosaccharyltransferase complex (OST) and a number of molecular chaperones and folding factors (BiP, calnexin, calreticulin, ERp57 and PDI).

framework on which to hang the molecular detail of the translocation process. In this respect, a major question that remains unresolved is what provides the actual driving force for the movement of the polypeptide chain through the ER translocon? Trying to understand the regulated opening and closing of the translocon ('gating'), both at its two ends and within the plane of the membrane, will be a major focus of future studies. An even greater challenge will be trying to fathom how the translocation site can accommodate so many accessory protein complexes and understanding how the structure and composition of the translocon alter during the translocation process.

Summary

- *SRP-dependent and SRP-independent targeting routes deliver precursor proteins to the ER membrane translocon.*
- *These precursors are translocated into (for membrane proteins) and across (for secretory protein) the ER membrane via aqueous channels composed of oligomers of the Sec61 complex.*
- *Both ends of the ER translocon are 'gated' and the opening and closing of these gates are closely regulated. The lateral exit of hydrophobic polypeptide regions into the phospholipid bilayer also appears to be a carefully controlled process.*
- *Accessory components are transiently associated with active ER translocation sites and modify the nascent polypeptide as it appears on the luminal side of the membrane.*

Limited space has meant that we have had to refer to much original source material via reviews. Our own research is supported by funding from the Biotechnology and Biological Sciences Research Council, the Medical Research Council and the Wellcome Trust.

References

1. High, S. & Dobberstein, B. (1992) Mechanisms determining the transmembrane disposition of proteins. *Curr. Opin. Cell Biol.* **4**, 581–586
2. Kutay, U., Ahnert-Hilger, G., Hartmann, E., Wiedenmann, B. & Rapoport, T.A. (1995) Transport route for synaptobrevin via a novel pathway of insertion into the endoplasmic reticulum membrane. *EMBO J.* **14**, 217–223
3. Rapoport, T.A., Jungnickel, B. & Kutay, U. (1996) Protein transport across the eukaryotic endoplasmic reticulum and bacterial inner membranes. *Annu. Rev. Biochem.* **65**, 271–303
4. Walter, P. & Johnson, A.E. (1994) Signal sequence recognition and protein targeting to the ER membrane. *Annu. Rev. Cell Biol.* **10**, 87–119
5. Freymann, D.M., Keenan, R.J., Stroud, R.M. & Walter, P. (1997) Structure of the conserved GTPase domain of the signal recognition particle. *Nature (London)* **385**, 361–364
6. Keenan, R.J., Freymann, D.M., Walter, P. & Stroud, R.M. (1998) Crystal structure of the signal sequence binding subunit of the signal recognition particle. *Cell* **94**, 181–191
7. Raden, D. & Gilmore, R. (1998) SRP-dependent targeting of ribosomes to the rough ER in the absence and presence of NAC. *Mol. Biol. Cell* **9**, 117–130
8. Neuhof, A., Rolls, M.M., Jungnickel, B., Kalies, K.-U. & Rapoport, T.A. (1998) Binding of SRP gives ribosome/nascent chain complexes a competitive advantage in ER membrane interaction. *Mol. Biol. Cell* **9**, 103–115
9. Möller, I., Jung, M., Beatrix, B., Levy, R., Kreibich, G., Zimmermann, R., Wiedmann, M. & Lauring, B. (1998) A general mechanism for regulation of access to the translocon: competition for a membrane attachment site on ribosomes. *Proc. Natl. Acad. Sci. U.S.A.* **95**, 13425–13430
10. Ogg, S.C., Barz, W.P. & Walter, P. (1998) A functional GTPase domain, but not its transmembrane domain, is required for function of the SRP receptor beta-subunit. *J. Cell Biol.* **142**, 341–354
11. Montoya, G., Svensson, C., Luirink, J. & Sinning, I. (1997) Crystal structure of the NG domain from the signal recognition particle receptor FtsY. *Nature (London)* **385**, 365–368
12. Rapiejko, P.J. & Gilmore, R. (1997) Empty site forms of the SRP54 and SRα GTPases mediate targeting of ribosome-nascent chain complexes to the endoplasmic reticulum. *Cell* **89**, 703–713
13. Bacher, G., Lütcke, H., Jungnickel, B., Rapoport, T.A. & Dobberstein, B. (1996) Regulation by the ribosome of the GTPase of the SRP during protein targeting. *Nature (London)* **381**, 248–251
14. High, S. (1995) Protein translocation at the membrane of the endoplasmic reticulum. *Progr. Biophys. Mol. Biol.* **63**, 233–250
15. Zimmermann, R. (1998) The role of molecular chaperones in protein transport into the mammalian endoplasmic reticulum. *Biol. Chem.* **379**, 275–282
16. Ng, D.T.W., Brown, J.B. & Walter, P. (1996) Signal sequences specify the targeting route to the endoplasmic reticulum membrane. *J. Cell Biol.* **134**, 269–278
17. Wahlberg, J.M. & Speiss, M. (1997) Multiple determinants direct the orientation of signal-anchor proteins: the topogenic role of the hydrophobic signal domain. *J. Cell Biol.* **137**, 555–562
18. High, S. & Laird, V. (1997) Membrane protein biosynthesis – all sewn up? *Trends Cell Biol.* **7**, 206–209
19. High, S., Laird, V. & Oliver, J.D. (1997) The biosynthesis of membrane proteins at the endoplasmic reticulum, in *Membrane Protein Assembly* (von Heijne, G., ed.), pp. 119–133, R.G. Landes Company, Austin
20. Martoglio, B. & Dobberstein, B. (1996) Snapshots of membrane-translocating proteins. *Trends Cell Biol.* **6**, 142–147

21. Matlack, K.E.S., Mothes, W. & Rapoport, T.A. (1998) Protein translocation: tunnel vision. *Cell* **92**, 381–390
22. Beckman, R., Bubuck, D., Grassucci, R., Penczek, P., Verschoor, A., Blobel, G. & Frank, J. (1997) Alignment of conduits for the nascent polypeptide chain in the ribosome-Sec61 complex. *Science* **278**, 2123–2126
23. Simon, S.M. & Blobel, G. (1991) A protein-conducting channel in the endoplasmic reticulum. *Cell* **65**, 371–380
24. Mothes, W., Jungnickel, B., Brunner, J. & Rapoport, T.A. (1998) Signal sequence recognition in cotranslational translocation by protein components of the endoplasmic reticulum membrane. *J. Cell Biol.* **142**, 355–364
25. Kalies, K.-U., Rapoport, T.A. & Hartmann, E. (1998) The β subunit of the Sec61 complex facilitates cotranslational protein transport and interacts with the signal peptidase during translocation. *J. Cell Biol.* **141**, 887–894
26. Hedge, R.S., Voigt, S., Rapoport, T.A. & Lingappa, V.R. (1998) TRAM regulates the exposure of nascent secretory proteins to the cytosol during translocation into the ER. *Cell* **92**, 621–631
27. Siegel, V. (1997) Recognition of a transmembrane domain: another role for the ribosome? *Cell* **90**, 5–8
28. Hamman, B.D., Hendershot, L.M. & Johnson, A.E. (1998) BiP maintains the permeability barrier of the ER membrane by sealing the lumenal end of the translocon pore before and early in translocation. *Cell* **92**, 747–758
29. Suzuki, T., Yan, Q. & Lennarz, W.J. (1998) Complex, two-way traffic of molecules across the membrane of the endoplasmic reticulum. *J. Biol. Chem.* **273**, 10083–10086
30. Greenfield, J.J.A. & High, S. (1999) The Sec61 complex is located in both the ER and the ER-Golgi intermediate compartment. *J. Cell Sci.* **112**, 1477–1486
31. Andrews, D.W. & Johnson, A.E. (1996) The translocon; more than a hole in the ER membrane? *Trends Biochem. Sci.* **21**, 365–369

2

The immunological properties of endoplasmic reticulum chaperones: a conflict of interest?

Christopher V. Nicchitta[1] and Robyn C. Reed

Department of Cell Biology, Duke University Medical Center, Durham, NC 27710, U.S.A.

Introduction

In eukaryotic cells, protein trafficking is initiated early in synthesis, as nascent secretory and membrane polypeptide chains are selected for translocation into the endoplasmic reticulum (ER). In the ER, newly synthesized proteins undergo rapid, chaperone-assisted protein folding and assembly reactions. Once folding and assembly are complete, proteins exit the ER and traffic to their proper destinations in the cell. In mammalian cells, an additional level of complexity is interposed over these processes. Mammalian cells have evolved the means to exploit the compartmental segregation of the protein-folding process to detect intracellular pathogens and to monitor the polypeptide composition of the cell. As the means to this end, peptides arising from intracellular proteolysis are transported into the ER, where they associate with peptide-binding integral membrane proteins, the major histocompatability complex (MHC) class-I heavy chains. Peptide binding to nascent MHC class-I molecules is a component of the MHC heavy chain folding pathway, and yields a highly stable form of the molecule. Once assembled, MHC–peptide

[1]*To whom correspondence should be addressed (e-mail: C.Nicchitta@cellbio.duke.edu).*

complexes are transported to the cell surface for subsequent interaction with CD8+ T-lymphocytes. This interaction permits the immune system to distinguish between normal (self) antigens and tumour or viral (non-self) antigens. Thus a compartmentally segregated folding pathway is coupled with a peptide-generation and -transport process, so that the intracellular polypeptide composition, as a selected peptide repertoire, is displayed on the cell surface. Insights into the composite regulation of these two primary pathways has provided a fascinating example of the varied, and in many ways unexpected, functions of chaperones in mammalian cells. In this chapter, we summarize current views on the contribution of chaperones to protein folding in the ER and review recent findings on the peptide-binding behaviour of ER chaperones and the immunogenicity of ER-chaperone–peptide complexes. Lastly, we speculate on the interplay between the polypeptide- and peptide-binding properties of ER chaperones and the potential functional consequences such interactions may have on polypeptide and peptide traffic in the ER.

Protein folding in the ER

In eukaryotic cells, the folding and assembly of nascent secretory and integral membrane proteins is segregated to the ER and is intimately dependent upon the activity of a set of resident ER chaperones: calnexin, calreticulin, GRP170, GRP94, BiP, ERp72, ERp57 and protein disulphide isomerase (PDI). These proteins have multiple, and in some cases overlapping, functions that reflect the kinetic and structural requirements for folding of a given secretory or membrane protein (Table 1). Included in this list of functions are (i) a lectin activity that contributes to the regulation of nascent glycoprotein folding, as displayed by calnexin and calreticulin [1]; (ii) the capacity to interact with hydrophobic domains, as seen with the Hsp70 protein, BiP [2,3]; and (iii) thiol oxidoreductase activity, which is necessary for the formation and disruption of disulphide bonds, and is displayed by PDI, ERp72 and ERp57 [4–6].

Depending on the folding pathway of a given nascent chain, chaperone–nascent-chain interactions may involve any or all of the chaperones, acting sequentially or in concert, at various stages of the folding pathway. It is difficult to predict, *a priori*, which chaperone function is necessary for a given protein substrate to fold properly. Clearly, a variety of chaperone–nascent-chain interactions can occur, and the summary activity of these interactions allows protein folding and assembly to proceed with the efficiency necessary for proper protein function and cell growth.

Chaperone–nascent-chain interactions may also serve critical functions in the recognition of mutated and/or improperly folded nascent chains. In a process whose molecular details remain to a large extent unknown, misfolded or improperly assembled proteins become substrates for proteolytic degradation. The degradation of misfolded proteins occurs, in large part, in the

Table 1. Molecular chaperones of the mammalian ER

Chaperone	Molecular mass (kDa)	Binding specificity	Putative function
GRP94 (gp96, endoplasmin, ERp99)	94	Polypeptides, peptides, glycoproteins, Ca^{2+}	Interact with folding proteins, elicit CD8+ T-cell response
Calreticulin (calregulin)	46	Peptides, glycoproteins, Ca^{2+}	Interact with folding proteins, ER Ca^{2+} sink, elicit CD8+ T-cell response
Calnexin (p88, IP90)	90 (TM)	Polypeptides, peptides, glycoproteins, Ca^{2+}	Retain incompletely assembled proteins in the ER
BiP (GRP78)	78	Polypeptides and peptides, especially hydrophobic sequences	Mask hydrophobic regions to prevent aggregation and misfolding
PDI (ERp59)	57	Polypeptides, peptides, nascent chains: binds through peptide backbone	Thiol oxidoreductase: form and break disulphide bonds; may assist in folding
ERp72	72	Polypeptides, Ca^{2+}	Thiol oxidoreductase, possible cysteine protease
ERp57 (GRP58, ERp61, Q2, HIP-70, CPT)	57	Glycoproteins	Thiol oxidoreductase
GRP170	170	Polypeptides, peptides	Assist in protein folding or assembly, assist in translocation

cytosol, and thus there exists a transport pathway that allows movement of such proteins from the ER to the cytosol [7]. This process, a form of structural quality control, also appears to contribute to steady-state turnover of native proteins [8].

Peptide binding to ER chaperones

The ability to bind reversibly to select peptide sequences is a necessary component of ER chaperone function. The peptide-binding activity of BiP has been studied by both stochastic selection, using a randomized peptide library, and phage display [2,3]. In both studies, a clear preference for aliphatic residues was observed. Statistical analysis of the phage-display data allowed identification of an optimal consensus BiP-binding sequence comprised of a heptameric motif of alternating aromatic and aliphatic residues [3]. Given the conclusion that BiP-peptide substrate selection is driven primarily by the physical properties of a given substrate, rather than a defined sequence, however, it is perhaps best to limit conclusions to the finding that BiP displays a clear binding preference for hydrophobic domains. Because hydrophobic domains display a propensity for non-specific aggregation, and are predominant sites of protein–protein interaction, by binding to such domains BiP can function to suppress aggregation and assist oligomeric protein assembly.

As noted previously, relatively little is known regarding the peptide-binding specificity of the other ER chaperones. PDI displays a broad and promiscuous peptide-binding activity [9], and has been identified as the prominent binding partner of peptides translocated into the ER [10]. Using peptides containing a photoreactive cross-linker, numerous other luminal chaperones have been identified as peptide-binding proteins [11]. These studies have proved valuable in identifying the cohort of peptide-binding proteins present in the ER, and have provided needed insights into substrate selection by the various chaperones. Insight into the structural criteria by which peptide substrate selection is governed, however, awaits further study.

MHC class-I peptide assembly

MHC class-I molecules function to provide a peptide-based representation of the intracellular protein composition of the cell to the immune system. As illustrated schematically in Figure 1, the primary site of peptide production is the cytosol, where proteins are subject to degradation by the proteasome complex. Degradation by the proteasome, a process limited predominantly to ubiquitinated proteins, yields a diverse array of peptides, a subset of which are competent for class-I binding. The initial level of selection for peptide loading on to class-I molecules is exerted at the level of the transporter associated with antigen presentation (TAP) [12]. The TAP complex is a resident ER oligomeric membrane protein that functions to transport peptides, in an ATP-dependent manner, from the cytosol to the ER lumen. TAP-mediated transport displays a marked preference for peptides of 8–11 amino acids, a length that closely approximates the optimal length for binding to the peptide-binding pocket of class-I molecules. Coincident with the TAP-mediated transport of peptides into the ER, nascent class-I molecules undergo a series of chaperone-assisted folding stages culminating with the formation of a multi-protein complex containing the TAP transporters [13]. It remains uncertain whether physical interaction of nascent class-I molecules with the TAP transporter functions in the direct transfer of peptide substrates to heavy-chain molecules and/or serves a role in the structural maturation of the heavy chain. Nonetheless, at this late stage of structural maturation, peptide loading occurs, the MHC molecule undergoes the final stages of protein folding, and class-I–peptide complexes are

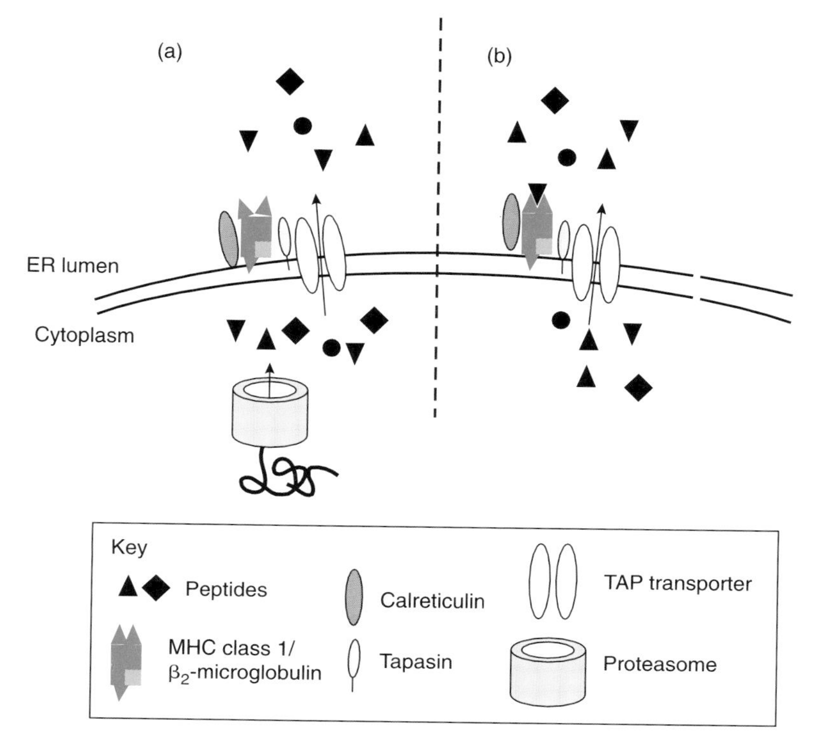

Figure 1. MHC-class-I–peptide complex formation in the ER
Prior to peptide binding, the folding MHC class-I molecule is stabilized in a complex with β_2-microglobulin, calreticulin, tapasin and TAP. Stable assembly of MHC class-I antigen requires a small peptide ligand. Such peptides are generated in the cytosol by proteasome-mediated protein degradation and transported into the ER lumen by the TAP transporter (a). Upon peptide binding to the MHC class-I molecule, full structural maturation is achieved. The complex disassembles, and the mature MHC class-I molecule is released from the ER for export to the cell surface and recognition by CD8+ T-cells (b).

transported to the cell surface for subsequent recognition by CD8+ cytotoxic T-lymphocytes.

ER chaperones as tumour-rejection antigens

In a pioneering study, Srivastava and colleagues [14] reported that immunization of mice with GRP94 derived from a chemically induced sarcoma elicited protective immunity to subsequent challenges with live tumour cells (e.g. Figure 2). Subsequent studies yielded a critical, and provocative, observation. That is, tumour-derived GRP94 was identical in sequence to the wild-type protein. Thus the observed immunogenic properties of GRP94 did not reflect the immunogenicity of the protein. To explain the ability of GRP94 to elicit a cellular immune response, Srivastava and colleagues [15] proposed that GRP94 immunogenicity was a consequence of a

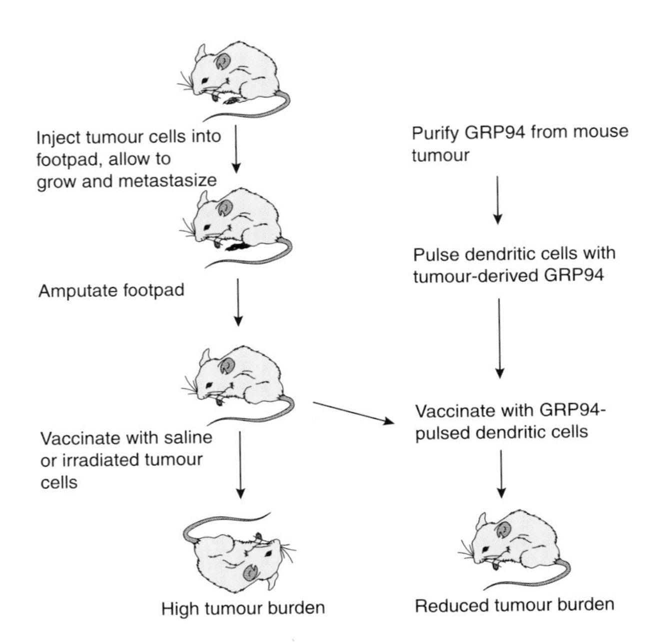

Figure 2. Vaccination with tumour-derived GRP94 or calreticulin elicits an immune response against the parent tumour
ER chaperones GRP94 and calreticulin are purified from tumour. Bone marrow-derived dendritic cells are pulsed *ex vivo* with the tumour-derived proteins. In parallel, mice are injected with tumour cells and subsequently the primary tumour mass is surgically removed. At this point, the mice carry a metastatic tumour burden. If vaccinated with dendritic cells pulsed with GRP94 or calreticulin derived from the same tumour line, the mice display a dramatically reduced rate of metastatic tumour progression.

population of stably bound peptides, at least a subset of which were unique to the parent tumour and were sufficient, when displayed on the class-I molecules of antigen-presenting cells, to elicit an immune response to the tumour. At the time that this hypothesis was proposed, it was unknown whether or not GRP94 was a peptide-binding protein and the argument for the role of bound peptides necessarily was rooted in analogy to known peptide-binding chaperones, such as BiP. Subsequent studies have confirmed and extended this observation, and it became clear that GRP94 co-purifies with a large host of stably bound host, bacterial and viral peptides [16,17]. The diversity of the GRP94-bound peptide pool became even more evident when it was reported that GRP94 co-purified with peptides whose transport into the ER was both TAP-dependent and TAP-independent [16–18]. On the basis of these studies it has been proposed that the array of GRP94-bound peptides is representative of the entire peptide repertoire of the ER [15].

The immunological significance of the peptide-binding activity of GRP94 was elevated further when it became clear that immunization with GPR94 elicited a CD8+ T-cell response to components of the bound-peptide pool [19,20]. Antigen presentation in the class-I pathway requires, with rare exception, the generation of peptides in the cytosol, their TAP-dependent transport into the ER, and subsequent loading on to nascent MHC class-I molecules. It can thus be presumed that elicitation of a CD8+ T-cell response by exogenous GRP94 requires it or its bound peptides to gain access to the ER of antigen-presenting cells. Only then, it is presumed, can the GRP94-derived peptides be loaded on to antigen-presenting-cell class-I molecules and CD8+ T-cells be activated. From an immunological perspective, the ability of GRP94 to gain access to the class-I-antigen presentation pathway from the extracellular space raises fascinating questions regarding the phenomenon of cross-priming [21]. For the purposes of this discussion, however, the observation that immunization with GRP94 elicits a CD8+ T-cell response is important because it indicates that (i) in the cell GRP94 stably binds peptide substrates and (ii) GRP94-bound peptides (or a subset thereof) can be loaded on to nascent class-I molecules.

The case for conflict

The ability of ER chaperones to bind both nascent polypeptides and peptides sets up a potential conflict of interest. Chaperones must reversibly bind nascent polypeptides to ensure their correct folding, but a constant influx of peptides intended for antigen presentation may compete for the same binding sites, resulting in a continuous kinetic battle of opposing ends and common means (Figure 3). If BiP can be used as a model for the behaviour of ER chaperones, it is established that peptides compete with unfolded proteins for access to the BiP polypeptide-binding pocket [22]. For BiP it is clear that the on and off rates for peptide binding are markedly influenced by the nucleotide-bound state of the protein, and thus there is a mechanistic basis for rapid exchange of peptide and polypeptide substrates. However, for GRP94 and calreticulin, the two ER chaperones documented to elicit CD8+ T-cell responses, peptide binding is quite stable. Although low-affinity interactions of GRP94 with ATP and ADP have been reported [23], it is not yet established how or if such interactions influence peptide binding to GRP94. In fact, studies to date indicate that adenine nucleotides are without effect on peptide binding to GRP94 [24]. That peptide binding to GRP94 and calreticulin is kinetically stable can be inferred readily from observations that the proteins retain their immunogenic activity throughout varied isolation and purification procedures and require denaturing conditions to induce peptide release *in vitro*.

Do such stable interactions with peptides influence the ability of GRP94 and calreticulin to function as chaperones? At present it is not known how GRP94 contributes to protein folding in the ER, although it is often found in association with structurally immature proteins. It is possible that GRP94 contains multiple domains for interaction with peptide and polypeptide substrates, and that such domains may function independently. Such a scenario has been

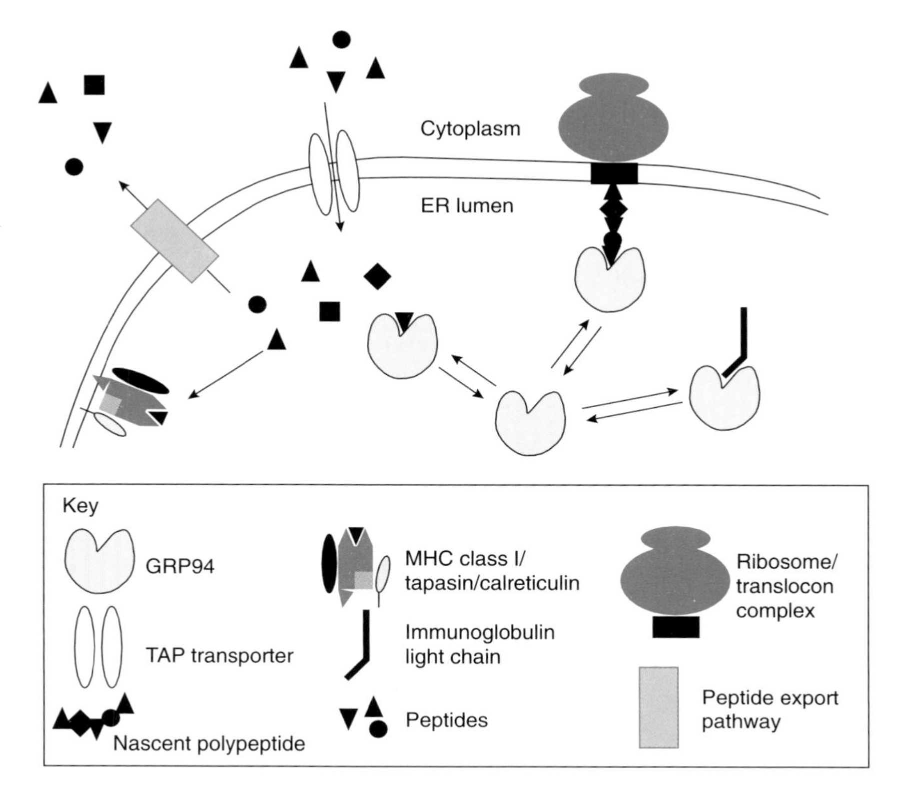

Figure 3. A scenario for multiple, competing functions for ER chaperones such as GRP94

GRP94 binds small peptides transported into the ER, as well as nascent chains and folding proteins such as immunoglobulin heavy and light chains. Assuming that these functions are all served by the same or overlapping domains of the protein, a competition for GRP94-binding sites among peptides, polypeptides and nascent chains is depicted. Peptides in the ER have several potential fates, including chaperone binding, MHC class-I antigen binding, and export by the peptide-export pathway.

proposed for the cytosolic paralogue of GRP94, Hsp90. It is not, however, clear how a domain could be selective for peptides, or polypeptides, only. One possibility is that, much in the manner of a class-I molecule, GRP94 contains a peptide-binding pocket that is 'closed' at both ends and thus can only accommodate peptides with discrete and free termini. If the postulate of a closed peptide-binding pocket is accepted, it remains difficult to explain how a polypeptide-binding site would discriminate against free peptides, without constraining binding selectivity to particular secondary structure characteristics of folding nascent chains. Alternatively, again in analogy to Hsp90, GRP94 may function as a chaperone scaffold and thereby physically assemble an array of chaperones and accessory factors [25]. Finally, since GRP94 can chaperone unfolded proteins in a ligand-regulated fashion *in vitro* [26], a ligand may regulate *in vivo* peptide- and/or polypeptide-binding activity.

For calreticulin, such an apparent conflict can be more easily rationalized. Assuming that the predominant chaperone function of calreticulin is expressed through its lectin-based interactions with the glycan moiety of nascent glycoproteins, a model in which calreticulin displays separate glycan and peptide binding domains would readily accommodate the multi-functional character of this protein. In this manner, calreticulin would be similar to PDI, which is both a peptide-binding protein and a thiol oxidoreductase.

Given the relative paucity of information on the contribution(s) that GRP94 makes to protein folding and assembly, this discussion is by nature conjecture. Whether stable peptide binding to GRP94 does, in fact, conflict with its proposed function as a chaperone is dependent on a number of factors. Included in such a list would be: (i) whether the peptide- and polypeptide-binding sites (if any) are shared; (ii) the relative peptide occupancy state of GRP94 at steady state; (iii) the lifetimes of GRP94–peptide complexes *in vivo*; (iv) the manner in which peptide binding and release is regulated; (v) the steady-state turnover rate of GRP94 and (vi) the presence of free peptides in the ER.

Summary

- *ER chaperones are abundant and highly conserved proteins that display both peptide binding and chaperone activity.*
- *Of the family of chaperones present in the mammalian ER, GRP94 and calreticulin are apparently unique in their ability to elicit CD8+ T-cell responses against components of their bound-peptide pools.*
- *The ability of GRP94 and calreticulin to elicit CD8+ T-cell responses indicates that both proteins bind peptides suitable for assembly on to MHC class-I molecules.*
- *The capacity to function as molecular chaperones and as peptide-binding proteins capable of transferring, directly or indirectly, peptides on to class-I molecules, indicates that GRP94 and calreticulin participate in the regulation of both peptide and polypeptide traffic in the ER.*
- *Perspectives on the regulation of and interplay between the peptide binding and chaperone activity of GRP94 and calreticulin are discussed.*

References

1. Helenius, A. (1994) How N-linked oligosaccharides affect glycoprotein folding in the endoplasmic reticulum. *Mol. Biol. Cell.* **5**, 253–265
2. Flynn, G.C., Pohl, J., Flocco, M.T. & Rothman, J.E. (1991) Peptide-binding specificity of the molecular chaperone BiP. *Nature (London)* **353**, 726–730
3. Blond-Elguindi, S., Cwirla, S., Dower, W.J., Lipshutz, R.J., Sprang, S.R., Sambrook, J.F. & Gething, M.-J.H. (1993) Affinity panning of a library of peptides displayed on bacteriophages reveals the binding specifity of BiP. *Cell* **75**, 717–728

4. Freedman, R.B. (1989) Protein disulfide isomerase: multiple roles in the modification of nascent secretory proteins. *Cell* **57**, 1069–1072
5. Hughes, E.A. and Cresswell, P. (1998) The thiol oxidoreductase ERp57 is a component of the MHC class I peptide-loading complex. *Curr. Biol.* **8**, 709–712
6. Mazzarella, R.A., Srinivasan, M., Haugejorden, S.M. & Green, M. (1990) ERp72, an abundant luminal endoplasmic reticulum protein, contains three copies of the active site sequences of protein disulfide isomerase. *J. Biol. Chem.* **265**, 1094–1101
7. Werner, E.D., Brodsky, J.L. & McCracken, A.A. (1996) Proteasome-dependent endoplasmic reticulum-associated protein degradation: an unconventional route to a familiar fate. *Proc. Natl. Acad. Sci. U.S.A.* **93**, 13797–13801
8. Kopito, R.R. (1997) ER quality control: the cytoplasmic connection. *Cell* **88**, 427–430
9. Noiva, R., Kimura, H., Roos, J. & Lennarz, W.J. (1991) Peptide binding by protein disulfide isomerase, a resident protein of the endoplasmic reticulum lumen. *J. Biol. Chem.* **266**, 19645–19649
10. Lammert, E., Stevanovic, S., Rammensee, H.-G. & Schild, H. (1997) Protein disulfide isomerase is the dominant acceptor for peptides translocated into the endoplasmic reticulum. *Eur. J. Immunol.* **27**, 1685–1690
11. Spee, P. and Neefjes, J. (1997) TAP-translocated peptides specifically bind proteins in the endoplasmic reticulum, including gp96, protein disulfide isomerase and calreticulin. *Eur. J. Immunol.* **27**, 2441–2449
12. Marusina, K. and Monaco, J.J. (1996) Peptide transport in antigen presentation. *Curr. Opin. Hematol.* **3**, 19–26
13. Sadasivan, B., Lehner, P.J., Ortmann, B., Spies, T. & Cresswell, P. (1996) Roles for calreticulin and a novel glycoprotein, tapasin, in the interaction of MHC class I molecules with TAP. *Cell* **5**, 103–114
14. Srivastava, P.K., DeLeo, A.B. & Old, L.J. (1986) Tumor rejection antigens of chemically induced tumors of inbred mice. *Proc. Natl. Acad. Sci. U.S.A.* **83**, 3407–3411
15. Srivastava, P.K., Udono, H., Blachere, N.E. & Li, Z. (1994) Heat shock proteins transfer peptides during antigen processing and CTL priming. *Immunogenetics* **39**, 93–98
16. Arnold, D., Wahl, C., Faath, S., Rammensee, H.-G. & Schild, H. (1997) Influences of transporter associated with antigen processing (TAP) on the repertoire of peptides associated with the endoplasmic reticulum-resident stress protein gp96. *J. Exp. Med.* **186**, 461–466
17. Srivastava, P.K., Menoret, A., Basu, S., Binder, R.J. & McQuade, K.L. (1998) Heat shock proteins come of age: primitive functions acquire new roles in an adaptive world. *Immunity* **8**, 657–665
18. Arnold, D., Faath, S., Rammensee, H.-G. & Schild, H. (1995) Cross-priming of minor histocompatability antigen-specific cytotoxic T cells upon immunization with the heat shock protein gp96. *J. Exp. Med.* **182**, 885–889
19. Udono, H., Levy, D.L. & Srivastava, P.K. (1994) Cellular requirements for tumor-specific immunity elicited by heat shock proteins: tumor rejection antigen gp96 primes CD8+ T cells *in vivo*. *Proc. Natl. Acad. Sci. U.S.A.* **91**, 3077–3081
20. Nicchitta, C.V. (1998) Biochemical, cell biological, and immunological issues surrounding the endoplasmic reticulum chaperone GRP94/gp96. *Curr. Opin. Immunol.* **10**, 103–109
21. Carbone, F.R., Kurts, C., Bennett, S.R.M., Miller, J.F.A.P. & Heath, W.R. (1998) Cross-presentation: a general mechanism for CTL immunity and tolerance. *Immunol. Today* **19**, 103–109
22. Fourie, A.M., Sambrook, J.F. & Gething, M.-J.H. (1994) Common and divergent peptide binding specificities of hsp70 molecular chaperones. *J. Biol. Chem.* **269**, 30470–30478
23. Rosser, M.F.N. & Nicchitta, C.V. (2000) Ligand interactions in the adenosine nucleotide binding domain of the Hsp90 chaperone, GRP94. I. Evidence for allosteric regulation of ligand binding. *J. Biol. Chem.* **275**, 22798–22805
24. Wearsch, P.A. and Nicchitta, C.V. (1997) Interaction of endoplasmic reticulum chaperone GRP94 with peptide substrates is adenine nucleotide-independent. *J. Biol. Chem.* **272**, 5152–5156

25. Csermely, P., Schnaider, T., Söti, C., Prohaszha, Z. and Nardi, G. (1998) The 90kDa molecular chaperone family: structure, function and clinical applications. A comprehensive review. *Pharmacol. Ther.* **79**, 1–39
26. Wassenberg, J.J., Reed, R.C. & Nicchitta, C.V. (2000) Ligand interactions in the adenosine nucleotide binding domain of the Hsp90 chaperone, GRP94. II. Ligand-mediated activation of GRP94 molecular chaperone and peptide binding activity. *J. Biol. Chem.* **275**, 22806–22814

3

Glycosylation and protein transport

Peter Scheiffele*[1] and Joachim Füllekrug†

**Department of Biochemistry and Biophysics, Howard Hughes Medical Institute, University of California, San Francisco, CA 94143-0452, U.S.A., †Max-Planck Institute for Molecular Cell Biology and Genetics, D-01307 Dresden, Germany*

Introduction

Glycans have a multitude of functions in organisms. They can be found conjugated to proteins, most commonly via a glycosidic linkage between the carbohydrate and the side chains of asparagine (N-glycans) or serine/threonine residues (O-glycans). As large numbers of different monosaccharides are incorporated into linear or branched structures, an enormously complex repertoire of carbohydrates can be created with far more permutations as observed for nucleic acids or polypeptides.

Why does the cell create this overwhelming diversity of structures? Studies performed during the last 20 years have revealed two principally different roles for protein-bound glycans: specific carbohydrate epitopes can serve as ligands for receptors that mediate recognition events, as adhesion between select cells. On the other hand, rather general glycan structures can be employed to change such biophysical properties of a protein as its charge, solubility, folding or sensitivity towards proteases [1]. Inhibition or deletion of glycosyltransferases normally does not influence growth of isolated cells, but embryogenesis is severely impaired. This highlights the importance of glycan structures in the multicellular organism.

[1]*To whom correspondence should be addressed (e-mail: Scheiffe@uclink4.berkeley.edu).*

Intracellularly, a specific carbohydrate epitope on lysosomal enzymes, the mannose 6-phosphate modification, has been shown to act as a signal for their delivery to lysosomes [2]. Besides this well-characterized sorting determinant, carbohydrates were thought to be primarily of structural importance for the folding of glycoproteins. However, recent studies suggested that carbohydrates can also provide information for other sorting events in the secretory pathway and the discovery of a new class of intracellular lectins has made this hypothesis even more attractive.

In this chapter we will discuss the role of glycans as sorting signals in the secretory pathway, specifically for exit of cargo molecules from the endoplasmic reticulum (ER) and the Golgi complex.

Transport to the cell surface by default?

In the ER, proteins with consensus sequences for N-glycosylation all acquire an identical large carbohydrate structure that is transferred 'en block' to the amido group of an asparagine. While passing through the secretory pathway this structure is then trimmed extensively and modified by glycosidases and glycosyltransferases to yield the large variety of N-glycans found in different glycoproteins. The carbohydrate-modification enzymes display a polarized localization throughout the Golgi stack, with the early acting enzymes being active in the *cis* and *medial* cisternae, and the late-acting enzymes in the *trans*-Golgi and *trans*-Golgi network.

According to the 'bulk-flow' hypothesis, proteins are transported passively on their route to the plasma membrane of mammalian cells, and early experiments suggested that carbohydrate modifications are not required for transport along the secretory pathway [3]. However, more recently elegant work on the calnexin/calreticulin quality-control machinery has demonstrated that the sequential processing of carbohydrates in the ER co-ordinates folding and exit of newly synthesized glycoproteins from the compartment [4]. Furthermore, evidence has accumulated indicating that efficient transport along the secretory pathway does require specific signals. Instead of being transported passively to the cell surface, cargo molecules were shown to be concentrated actively in the transport carriers [5]. The identification of new lectins and the finding that carbohydrates can act as intracellular sorting signals suggests that carbohydrates might play an important role in cargo selection in the secretory pathway (see Figure 1).

Lectins in the secretory pathway

Lectins are proteins that selectively bind to specific carbohydrate structures. Recently, a novel class of animal lectins, homologous to leguminous-plant lectins, has been identified. Although the overall homology of this protein family to legume lectins is low, the amino acids constituting the carbohydrate-binding site are conserved [6,7].

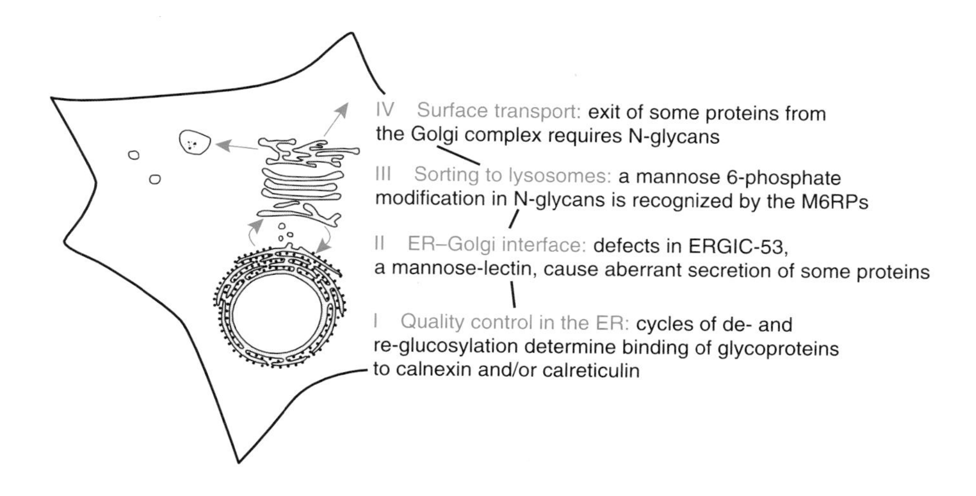

Figure 1. Involvement of lectins or carbohydrates in different steps of the secretory pathway
See text for details. M6PR, mannose-6-phosphate receptor.

ERGIC-53 (ER-Golgi intermediate-compartment protein of 53 kDa) is the most extensively characterized member of this new class of lectins. It binds specifically to mannose-containing carbohydrate structures in a calcium-dependent manner. This interaction is abolished by mutation of amino acid residues in the sugar-binding site, as predicted by sequence comparison with leguminous lectins [8].

Human individuals deficient in ERGIC-53 suffer from a bleeding disorder, a rare autosomal recessive disease referred to as combined deficiency of coagulation factors V and VIII [9]. Both coagulation factors are heavily glycosylated and are reduced to 5–30% of the normal levels in plasma. Other than that, these patients do not show any other obvious symptoms, implying that embryonic development and secretion of other glycoproteins are not seriously affected. This is somewhat surprising since ERGIC-53 is an abundant protein and widely expressed in different tissues, suggesting a housekeeping function. However, it seems likely that ERGIC-53 enhances the efficiency of secretion in general, and other functionally redundant factors might be able to compensate for the lack of ERGIC-53 during development.

ERGIC-53 has been localized to the ER, ER-Golgi intermediate compartment and *cis*-Golgi in mammalian tissue-culture cells. It is an integral membrane protein that spans the membrane once, with its N-terminus localized to the lumen of the membrane compartment (a type-I membrane protein). A short cytoplasmic tail contains several determinants for sorting in the early secretory pathway. A double phenylalanine motif at the C-terminus has been shown to interact with cytoplasmic coat proteins (COP II), which are involved in budding of transport vesicles from the ER. These coat proteins select cargo molecules (see Chapter 4 in this volume by Francis Barr) and an attractive hypothesis suggests that a certain class of membrane proteins would interact with COP II on the cytoplasmic side and with secretory cargo proteins in the lumen of the ER. Such putative coat-cargo receptors would thus provide a bridge through the membrane bilayer, extending the cargo-selection capabilities of the COP II coat. The cytoplasmic tail of ERGIC-53 also contains a dilysine ER localization motif directly interacting with coatomer (COP I) [8]. COP I proteins are thought to bind to the cytoplasmic tail of proteins in the Golgi apparatus and guide their retrieval to the ER. The combination of anterograde (COP II) and retrograde (COP I) signals is assumed to confer cycling of ERGIC-53 between ER and *cis*-Golgi, leading to a steady-state localization in the ER-Golgi intermediate compartment.

In view of all these properties, ERGIC-53 has been suggested to constitute a cargo receptor which would concentrate glycoproteins at ER exit sites. Following transport to the Golgi apparatus, changes in calcium concentration and/or pH would release the glycoproteins, and enable ERGIC-53 to recycle back to the ER to collect a new set of glycosylated secretory proteins. Alternatively, ERGIC-53 could be involved in quality control in the early secretory pathway. Incorrectly glycosylated proteins would be either retained or retrieved until they were modified properly. Interestingly, extensive trimming of mannose residues occurs at the ER–Golgi interface, possibly modifying the mannose-containing epitope recognized by ERGIC-53.

Expression of a mutant of ERGIC-53 that localizes exclusively to the ER (but retains lectin activity) results in impaired secretion of at least one lysosomal enzyme. However, other glycoproteins tested were not affected. Therefore, not all secretory proteins depend on the recycling of ERGIC-53 for efficient secretion. Still, these experiments prove the point that *cycling* of this protein is functionally relevant [8].

In mammalian cells, another abundant type-I membrane protein related to ERGIC-53 has been described. First identified in epithelial cells, VIP36 (vesicular integral membrane protein of 36 kDa) also displays significant sequence homology to leguminous-plant lectins. Similar to ERGIC-53, VIP36 recognizes a high-mannose sugar structure on glycoproteins and localizes to the early secretory pathway [10,11]. Future experiments will be required to clarify whether VIP36 and ERGIC-53 have common ligands and whether they are functionally redundant or complementary.

Glycosylation-dependent sorting of proteins in the secretory pathway

Studies performed in 1985 suggested that N-glycosylation might be required for transport of transmembrane proteins to the cell surface. Rose and co-workers [12] observed that a non-glycosylated membrane-anchored secretory protein (rat growth hormone) was efficiently transported to the Golgi

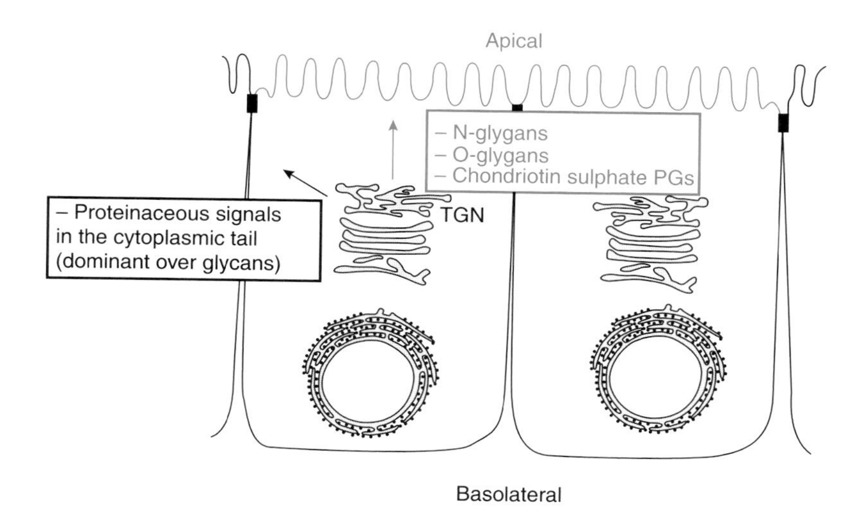

Figure 2. Sorting in the *trans*-Golgi network of polarized epithelial cells
In MDCK cells N-glycans can act as apical sorting signals for secretory and transmembrane proteins. For the $p75^{NTR}$ the O-glycosylated stalk is required for apical sorting and analysis of secreted proteoglycans revealed that chondroitin sulphate is apically secreted. Basolateral signals are found in the cytoplasmic tail of transmembrane proteins and can override an apical signal. Secretory proteins without signals are secreted in a non-polarized fashion and transmembrane proteins without signals are retained in the Golgi complex. TGN, *trans*-Golgi network; PG, proteoglycan.

complex, but did not reach the cell surface. Introduction of N-glycosylation sites by site-directed mutagenesis allowed exit from the Golgi and subsequent transport to the cell surface. Similarly, surface transport of the vesicular stomatitis virus glycoprotein (VSVG) required the presence of at least one N-linked carbohydrate chain. However, N-glycosylation is not required for exocytosis of all proteins.

Some new insight was gained from the analysis of protein sorting in polarized epithelial cells (see Figure 2). Apical sorting of the major secretory protein in Madin–Darby canine kidney (MDCK) cells, gp80, was lost when cells were treated with tunicamycin, an inhibitor of N-glycosylation [13]. Furthermore, addition of N-glycosylation sites into the normally non-glycosylated and non-polarized secreted rat growth hormone was sufficient to target the protein apically. Interestingly, the number of N-glycosylation sites introduced correlated with the efficiency of apical sorting [14].

This function of N-glycans in sorting is not restricted to epithelial cells. In fibroblasts, neurons and osteoclasts at least two different post-Golgi routes to the cell surface operate [15]. These routes use mechanisms similar to the apical and basolateral pathways and have therefore been termed apical and basolateral 'cognate' routes. Using three different model proteins, Matter and colleagues [16] demonstrated a role for N-glycosylation in surface transport in fibroblasts. In all cases the unglycosylated forms of the analysed proteins accumulated in the Golgi complex, whereas N-glycosylation allowed surface delivery. In MDCK cells the same glycosylated proteins were sorted to the apical membrane, suggesting that also for membrane proteins N-glycans can act as an apical signal [16]. Basolateral sorting signals are generally found in the cytoplasmic tail of membrane proteins. When such a cytoplasmic signal was added to any of the three model proteins they were targeted basolaterally, independently of N-glycosylation. Importantly, also in fibroblasts addition of a basolateral signal allowed carbohydrate-independent surface delivery. The simplest explanation for these results is that exit from the Golgi complex is signal-mediated. According to this hypothesis, transport of membrane proteins to the cell surface requires sorting determinants to direct cargo into either the apical or the basolateral (cognate) carriers. As suggested by previous studies the cytoplasmic basolateral signals are generally dominant over the apical information [17]. However, in case a protein carries no basolateral signal then positive apical information (as N-glycosylation) is required for exit from the Golgi complex.

Clearly, N-glycans can act as determinants of apical sorting; however, the presence of N-glycans in a molecule does not guarantee that the protein will be transported apically. Deletion of the basolateral signal in the cytoplasmic tail of the transferrin receptor results in non-polarized delivery, although the protein contains two N-linked glycans. Most likely, the positioning of carbohydrates on the surface of the molecule is critical, but also additional sorting information in the transmembrane domain can be required for apical delivery [18]. In the case of the neurotrophin receptor ($p75^{NTR}$) deletion of the single N-glycosylation site did not abolish apical delivery, instead the juxtamembrane domain, which is modified by O-glycosylation, was required for sorting [19]. Direct evidence for a role of O-glycosylation in apical targeting is still lacking, although there are other examples of O-glycosylated proteins that are apically sorted in the absence of N-glycans.

Besides N- and O-glycans, secretory proteins can also be modified by addition of proteoglycan chains. Interestingly, a differential sorting of secreted proteoglycans in polarized cells was observed recently. Whereas only 20% of heparan sulphate proteoglycans are secreted apically, most of the chondroitin sulphate proteoglycans (75%) are secreted into the apical medium [20]. Heparan sulphate and chondroitin sulphate glycosaminoglycan chains differ mainly in their repeating disaccharides. Since protein-free chondroitin sulphate chains are also secreted predominantly to the apical medium, the sorting information must be localized in the sugar chains. However, the precise nature of this sorting signal remains to be elucidated.

How is sorting information in glycans decoded?

While there is now strong evidence for carbohydrate signals in the secretory pathway, it is far less clear how all these different sorting determinants are interpreted. The mannose 6-phosphate epitope in the lysosomal enzymes is recognized by its receptor which has lectin activity. Are there more intracellular lectins recognizing sugar structures in N-glycans, O-glycans and proteoglycans in the apical post-Golgi route? Another possibility would be that instead of acting as specific epitopes the glycans could change general properties of proteins and for example drive self-association. Clustering of cargo molecules into targeting patches with specific lipid composition, so-called rafts, has been proposed to mediate apical sorting [21]. This clustering could be driven by lectins associated with raft domains, which display low-affinity interactions with carbohydrates of the cargo proteins. However, another possibility is that the glycan structures on the cargo molecules themselves mediate co-aggregation under the chemical conditions of the *trans*-Golgi network, similar to that which has been proposed for sorting of proteins into secretory granules for regulated secretion [22].

A requirement for specific terminal carbohydrate modifications in apical sorting has been tested by treatment of tissue-culture cells with pharmacological inhibitors of terminal glycosylation. In MDCK cells deoxymannojirimycine or swainsonine, which both inhibit the formation of complex-type N-glycans, did not disturb apical sorting of gp80 [23]. In HT-29 cells, however, inhibition of Galβ1-3GalNAc:2,3-sialyltransferase (and thereby terminal sialylation) resulted in an intracellular accumulation of several apical proteins whereas basolateral delivery was not affected [24]. It is possible that in the mucus-producing HT-29 cells signals additional to those found in MDCK cells operate. However, clearly further studies and the identification of apical sorting receptors will be required to define precisely the mechanism of apical sorting directed by carbohydrate signals.

Perspectives

Only a few years ago, the role of glycosylation in the secretory pathway was thought to be limited to sorting of lysosomal proteins by mannose-6-phosphate receptors. Since then, surprising and exciting progress has unveiled a role for glycosylation in ER quality control and generation of epithelial polarity at the level of the *trans*-Golgi network. In addition, a new class of lectins localized to the Golgi apparatus has been discovered, which still waits for a firm functional assignment.

Nevertheless, these intracellular leguminous lectin homologues have been one of the pieces of evidence which led to the general idea that protein transport in the secretory pathway does not occur by default, but is signal-mediated. The human proteins ERGIC-53 and VIP36 would be engaged in lectin–glycoprotein interactions, concentrating and chaperoning glycoproteins on their route along the secretory pathway, or proof-reading their carbohydrate modifications. Sequence database searches reveal ERGIC-53 and VIP36 homologues in eukaryotic organisms as diverse as yeast, worms and fruitflies, suggesting evolutionarily conserved roles in the exocytic pathway.

The multitude of different carbohydrate signals operating in post-Golgi trafficking makes it entirely possible that whole families of specific intracellular lectins have not yet been identified. However, an alternative explanation is that those signals act rather by driving co-aggregation of cargo molecules or association with few lectin-like receptors of very low carbohydrate specificity. Although there are still many open questions it now seems likely that in every eukaryotic cell exocytosis requires signals that target cargo on specific post-Golgi routes to the cell surface.

Summary

- *Transport along the secretory pathway is largely signal-mediated.*
- *Proteins in the secretory pathway can be covalently modified with various carbohydrate structures, most commonly O-glycans, N-glycans and/or proteoglycans.*
- *Carbohydrate modifications can change the physical properties of proteins or can function as specific recognition epitopes.*
- *Glycosylation can act as an apical sorting signal in polarized epithelial cells and provide a signal for surface transport in non-polarized fibroblasts.*
- *Homologues of leguminous plant lectins have been identified in yeast, fruitflies, worms and humans.*
- *Intracellular lectins are candidate receptors in the secretory pathway to mediate concentration of cargo in carrier vesicles.*

We are grateful for financial support from the German research foundation (Deutsche Forschungsgemeinschaft), the Max-Planck Society and the European Molecular Biology Organization.

References

1. Varki, A. (1993) Biological roles of oligosaccharides: all of the theories are correct. *Glycobiology* **3**, 97–130
2. Kornfeld, S. (1992) Structure and function of the mannose 6-phosphate/insulin-like growth factor II receptors. *Annu. Rev. Biochem.* **61**, 307–330
3. Wieland, F.T., Gleason, M.L., Serafini, T.A. & Rothman, J.E. (1987) The rate of bulk flow from the endoplasmic reticulum to the cell surface. *Cell* **50**, 289–300
4. Trombetta, E.S. & Helenius, A. (1998) Lectins as chaperones in glycoprotein folding. *Curr. Opin. Struct. Biol.* **8**, 587–592
5. Balch, W.E., McCaffery, J.M., Plutner, H. & Farquhar, M.G. (1994) Vesicular stomatitis virus glycoprotein is sorted and concentrated during export from the endoplasmic reticulum. *Cell* **76**, 841–852
6. Sharon, N. & Lis, H. (1990) Legume lectins – a large family of homologous proteins. *FASEB J.* **4**, 3198–3208

7. Fiedler, K. & Simons, K. (1994) A putative novel class of animal lectins in the secretory pathway homologous to leguminous lectins. *Cell* **77**, 625–626
8. Hauri, H.-P., Kappeler, F., Andersson, H. & Appenzeller, C. (2000) ERGIC-53 and traffic in the secretory pathway. *J. Cell Sci.* **113**, 587–596
9. Nichols, W.C., Seligsohn, U., Zivelin, A., Terry, V.H., Hertel, C.E., Wheatley, M.A., Moussalli, M.J., Hauri, H.P., Ciavarella, N., Kaufman, R.J. & Ginsburg, D. (1998) Mutations in the ER-Golgi intermediate compartment protein ERGIC-53 cause combined deficiency of coagulation factors V and VIII. *Cell* **93**, 61–70
10. Hara-Kuge, S., Ohkura, T., Seko, A. & Yamashita, K. (1999) Vesicular-integral membrane protein, VIP36, recognizes high-mannose type glycans containing alpha1→2 mannosyl residues in MDCK cells. *Glycobiology* **9**, 833–839
11. Füllekrug, J., Scheiffele, P. & Simons, K. (1999) VIP36 localisation to the early secretory pathway. *J. Cell Sci.* **112**, 2813–2821
12. Guan, J.-L., Machamer, C.E. & Rose, J.K. (1985) Glycosylation allows cell-surface transport of an anchored secretory protein. *Cell* **42**, 489–496
13. Urban, J., Parczyk, K., Leutz, A., Kayne, M. & Kondor-Koch, C. (1987) Constitutive apical secretion of an 80-kDa sulfated glycoprotein complex in the polarized epithelial Madin-Darby canine kidney cell line. *J. Cell Biol.* **105**, 2735–2743
14. Scheiffele, P., Peränen, J. & Simons, K. (1995) N-glycans as apical sorting signals in epithelial cells. *Nature (London)* **378**, 96–98
15. Keller, P. & Simons, K. (1997) Post-Golgi biosynthetic trafficking. *J. Cell Sci.* **110**, 3001–3009
16. Gut, A., Kappeler, F., Hyka, N., Balda, M.S., Hauri, H.-P. & Matter, K. (1998) Signal-mediated exit from the Golgi complex and apical targeting of membrane proteins. *EMBO J.* **17**, 1919–1929
17. Matter, K. & Mellman, I. (1994) Mechanisms of cell polarity: sorting and transport in epithelial cells. *Curr. Opin. Cell Biol.* **6**, 545–554
18. Kundu, A., Avalos, R.T., Sanderson, C.M. & Nayak, D.P. (1996) Transmembrane domain of influenza virus neuraminidase, a type II protein, possesses an apical sorting signal in polarized MDCK cells. *J. Virol.* **70**, 6508–6515
19. Yeaman, C., Le Gall, A.H., Baldwin, A.N., Monlauzeur, L., Le Bivic, A. & Rodriguez-Boulan, E. (1997) The O-glycosylated stalk domain is required for apical sorting of neurotrophin receptors in polarized MDCK cells. *J. Cell Biol.* **139**, 929–940
20. Kolset, S.O., Vuong, T.T. & Prydz, K. (1999) Apical secretion of chondroitin sulfate in polarized Madin-Darby canine kidney (MDCK) cells. *J. Cell Sci.* **112**, 1797–1801
21. Simons, K. & Ikonen, E. (1997) Sphingolipid-cholesterol rafts in membrane trafficking and signalling. *Nature (London)* **387**, 569–572
22. Thiele, C., Gerdes, H.-H. & Huttner, W.B. (1997) Protein secretion: puzzling receptors. *Curr. Biol.* **7**, R496–R500
23. Wagner, M., Morgans, C. & Koch-Brandt, C. (1995) The oligosaccharides have an essential but indirect role in sorting gp80 (clusterin, TRPM-2) to the apical surface of MDCK cells. *Eur. J. Cell. Biol./* **67**, 84–88
24. Huet, G., Hennebicq-Reig, S., de Bolos, C., Ulloa, F., Lesuffleur, T., Barbat, A., Carrière, V., Kim, I., Real, F.X., Delannoy, P. & Zweibaum, A. (1998) GalNAc-α-O-benzyl inhibits NeuAcα2-NeuAcα3 glycosylation and blocks the intracellular transport of apical glycoproteins and mucus in differentiated HT-29 cells. *J. Cell Biol.* **141**, 1311–1322

4

Vesicular transport

Francis Barr[1]

IBLS, Division of Biochemistry and Molecular Biology, University of Glasgow, CRC-Beatson Laboratories, Garscube Estate, Switchback Road, Glasgow G61 1BD

Introduction

In recent years there have been many advances in our understanding of the processes by which the vesicles of the secretory pathway are formed and then recognize and fuse with their target membrane. Possibly the most exciting advances in this area are the successful reconstitutions of vesicle formation from purified protein and lipid components [1–3]. The other area in which there have been many recent advances is in the mechanisms by which vesicles specifically recognize and fuse with their target membranes. This process has been divided into an early reversible event thought to occur over relatively long distances, termed 'tethering', and a late event occurring as the vesicle approaches its target membrane, termed 'docking' [4]. This chapter will focus on these advances, paying particular attention to the mechanisms involved in vesicle transport.

Coat proteins and secretory vesicle formation

There are four main classes of vesicle described in association with the secretory pathway. These are the coat promoter (COP) I and COP II vesicles involved in transport between the endoplasmic reticulum (ER) and the Golgi and within the Golgi, and the 'lace-like coat' and multiple types of clathrin-coated vesicles associated with the *trans*-Golgi network [5]. Since little is

[1]*Address for correspondence: Max Planck Institute for Biochemistry, Department of Cell Biology, Am Klopferspitz 18a, D-82152 Martinsreid, Germany*

known about the lace-like coat, and the subject of clathrin-coated vesicles together with the various clathrin adaptor proteins is a topic in its own right, this chapter will be restricted to COP I- and COP II-coated vesicles.

COP I-coated vesicle formation

After clathrin the second vesicle coat to be identified was the coatomer or COP I vesicle-coat protein complex. COP I-coated vesicles were first characterized in an *in vitro* assay for intra-Golgi transport [6,7]. COP I comprises seven polypeptide chains in a single globular complex of approx. 700 kDa, is found on vesicles derived from the Golgi apparatus, and requires the small GTP-binding protein Arf1p for its recruitment to membranes [6,7]. The role of Arf1p in coatomer recruitment has been explained in two ways, as a direct anchor for the coat complex [7], or indirectly via activation of phospholipase D [9]. Activation of phospholipase D results in the generation of phosphatidic acid by hydrolysis of phosphatidylcholine, and it has been shown that this can lead to the generation of a membrane surface able to bind COP I [9], leading to the hypothesis that Arf1 acts indirectly during COP I vesicle formation. The recent reconstitution of COP I vesicles from chemically defined liposomes *in vitro* has now allowed the role of Arf1p to be studied in detail, and the results obtained appear to support a direct role for it in coat recruitment [1], consistent with evidence that it can be directly cross-linked to the β-subunit of COP I [10]. In the presence of Arf1p and a non-hydrolysable analogue of GTP the COP I coat complex was able to bind to approx. 300 nm synthetic liposomes and to give rise to coated 40–70 nm vesicle-like structures [1]. These results are essentially the same as seen for Sar1p and COP II coats [3], and imply that a coat-protein complex and its corresponding small GTP-binding protein are all that is needed to deform a membrane and to carry out the scission event that produces a free vesicle (Figure 1b). There is also a potential role for cargo molecules in the recruitment of the COP I coat, since it can interact with the di-lysine type of ER retrieval motif found on a number of proteins recycling between the ER and Golgi apparatus [11].

Possibly the most controversial area in the secretory pathway at the present time is the role that COP I plays in transporting material through the Golgi apparatus. For many years it has been known that some secretory content is too large to fit into a small 70 nm coated vesicle such as the COP I vesicle [12]. This has lead to the suggestion that transport through the Golgi apparatus is not mediated by forward-moving transport vesicles but by a maturation process with new cisternae forming on the *cis*-face and older cisternae disassembling on the *trans*-face of the stack. This hypothesis suggests that the enzymes and other Golgi-resident proteins are continuously being recycled to earlier cisternae in the stack or to the ER by retrograde vesicles, proposed to be COP I vesicles. Direct evidence for maturation has until now been felt to be lacking, but a recent study investigating the secretion of procollagen fibres

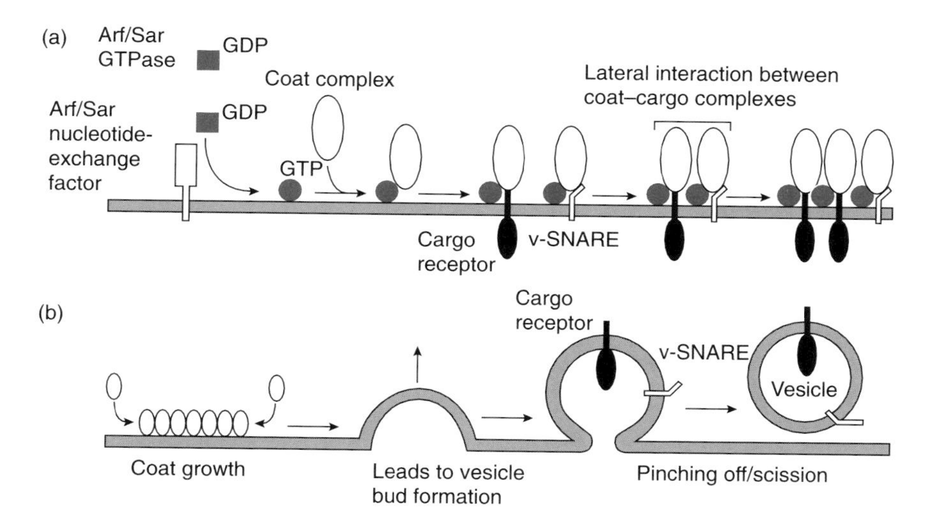

Figure 1. Coat nucleation and vesicle formation
(a) Coat nucleation. An exchange factor associated with the membrane triggers nucleotide exchange on a small GTP-binding protein of the Arf/Sar family. The GTP form of the Arf/Sar protein binds to the membrane recruiting the coat complex, which then interacts with cargo molecules. Coat–cargo complexes then start to interact laterally. (b) Vesicle budding. Lateral interaction of coat–cargo complexes leads to deformation of the membrane to form a vesicle bud. When the membrane has become sufficiently deformed a pinching-off/scission event occurs, leading to the release of a free vesicle. v-SNARE, vesicle soluble *N*-ethylmaleimide-sensitive fusion protein attachment protein receptor.

clearly demonstrates that these 300 nm-long fibres move through the Golgi apparatus without ever leaving the lumen of the cisternae [12]. However, there is also evidence that COP I vesicles may be able to transport material in both an anterograde and retrograde fashion through the Golgi apparatus [13], so it is possible that some secretory cargo it transported forwards by this means. Further work is obviously needed in this area to ascertain which cargo molecules move forward in maturing cisternae, what role forward-transport vesicles might play in cargo transport, and whether or not all Golgi enzymes are rapidly recycled via retrograde vesicles.

COP II-coated-vesicle formation

COP II vesicles were discovered following the identification of their coat components in a genetic screen for genes required for secretion in yeast [6]. These vesicles are thought to be involved in the transport of proteins from the ER to the early Golgi [6]. Their coat comprises three basic components, the small GTPase Sar1p, and two other complexes, Sec23/24p and Sec13/31p [6]. Sar1p binds to the ER membrane when in the GTP form and recruits the Sec23/24p complex, which subsequently binds the Sec13/31p complex [14]. These events have now been reconstituted in an *in vitro* system using liposomes with defined lipid composition [3]. In this system it is possible to produce 50–90 nm COP II-coated-vesicle-like structures with only the three components mentioned above and a non-hydrolysable analogue of GTP. These results imply that the binding of the COP II coat to the lipid bilayer is sufficient to mechanically deform it into a vesicle and to cause the final scission event resulting in the production of a free vesicle (Figure 1b). These results do not rule out that additional components are involved in this process in living cells, but do suggest there is no need for a dedicated scission protein or ‘pinchase’ such as dynamin, which has this function in endocytic vesicle formation from the plasma membrane. It should be noted that the ratio of Sar1p to other coat subunits is not the same in true COP II vesicles and those produced from liposomes, and that to obtain optimal formation of COP II vesicles from liposomes it is necessary to use 5-fold more of the coat subunits [3]. These observations might indicate that additional factors are important for stable coat interaction, although in their absence an excess of Sar1p is clearly able to substitute for them. One candidate factor is Sec16p, a protein localized to the ER membrane that is known to physically interact with Sec23p, Sec24p and Sec31p [15,16]. Other factors are likely to be the cargo molecules packaged into the COP II vesicles *in vivo*, since it has been shown that these are able to interact directly with the Sar1p-GTP–Sec23/24p complex [17].

Coat nucleation and cargo recruitment

Probably the biggest question remaining in the field of vesicle formation is how specific vesicle coats are recruited to the correct membrane to initiate the formation of a vesicle, a process termed coat nucleation (Figure 1a). Initiation of COP II binding to the ER membrane involves an ER membrane protein, Sec12p, that acts as a GTP–GDP exchange factor for Sar1p [6]. One set of proteins apparently important for coat nucleation are the vesicle proteins required for specific fusion of the vesicle with its target membrane (see below). Using purified recombinant proteins it has been demonstrated that Bos1p, which functions in the fusion of ER-derived vesicles with the Golgi, can bind to the Sar1p–Sec23/24p complex [18]. These observations are supported by experiments in which ER-resident and COP II-vesicle cargo proteins were expressed as soluble recombinant proteins lacking their transmembrane domains, and then immobilized on the surface of liposomes [19]. These liposomes were then exposed to the COP II coat components, and the amount of the different proteins packaged into COP II vesicles quantified. Whereas Bos1p, a membrane protein required for vesicle fusion, was concentrated in the COP II vesicles, the related ER-resident protein Ufe1p, and Sec12p, the exchange factor for Sar1p, were present at the same concentration as in the starting liposomes. Sec16p is currently the best candidate for a factor regulating coat nucleation, since it displays interactions with multiple components of the COP II coat [16]. It is possible that Sec16p cross-links multiple COP II-coat subunits interacting with cargo molecules into a 'polyvalent network' [19]. If this is the case, then Sec16p must release the coat subunits prior to the vesicle scission event, since like Sec12p it is not present in free COP II vesicles.

Factors controlling COP I vesicle nucleation are less well understood. These vesicles are known to be enriched in certain molecules being recycled from the Golgi apparatus to the ER, such as the KDEL receptor and ERGIC-53 (ER-Golgi intermediate-compartment protein of 53 kDa) [20]. Recently, direct evidence for the role of the p23 family of putative ER-to-Golgi cargo receptors has been obtained using a system reconstituting the formation of COP I vesicles from liposomes [2]. The cytoplasmic domain of one of the p23 proteins was linked to the surface of artificial liposomes, which were able to produce COP I-coated-vesicle-like structures when exposed to COP I, Arf1p and a non-hydrolysable analogue of GTP [2]. When the p23 cytoplasmic domain was omitted from the reaction, or its sequence was mutated to abolish the KKXX COP I-binding motif, vesicle formation was impaired. As previously discussed, it is also possible to make COP I vesicles from liposomes without such a molecule, but only when lipid mixes with non-physiological acyl-chains are used [1,2].

Target recognition

How does a vesicle find and recognize its target membrane from a distance many times its own diameter? One important aspect of this process is the vectorial transport of newly formed vesicles towards their target by the activity of specific motor proteins. Examples of this include the transport towards the Golgi apparatus of ER-derived transport intermediates along microtubules by kinesin II [21], and the transport of secretory vesicles to the daughter cell by class-V myosin II in budding yeast [22]. As a vesicle approaches its target membrane a recognition process is thought to occur involving the Rab proteins and members of a growing family of coiled-coil proteins involved in membrane traffic [23] (see Figure 2). This recognition process has been termed tethering and is upstream of the final docking and membrane-fusion events [4]. In transport between the ER and Golgi apparatus, the coiled-coil protein Uso1p has been shown to function upstream of docking mediated by SNARE [soluble *N*-ethylmaleimide-sensitive fusion protein (NSF) attachment protein receptor] proteins in yeast [24]. Using a yeast cell-free transport assay it has been shown that Uso1p action in tethering requires Ypt1p, a member of the Rab family of small GTPases, but is independent of the SNARE proteins [25]. The mammalian homologue of Uso1p, a protein named p115, has been shown to be able to tether COP I vesicles to Golgi membranes in an *in vitro* binding assay [26]. In this system the binding sites for p115 on the vesicle and target membranes are known, these are Giantin on the COP I vesicle and GM130 on the Golgi apparatus

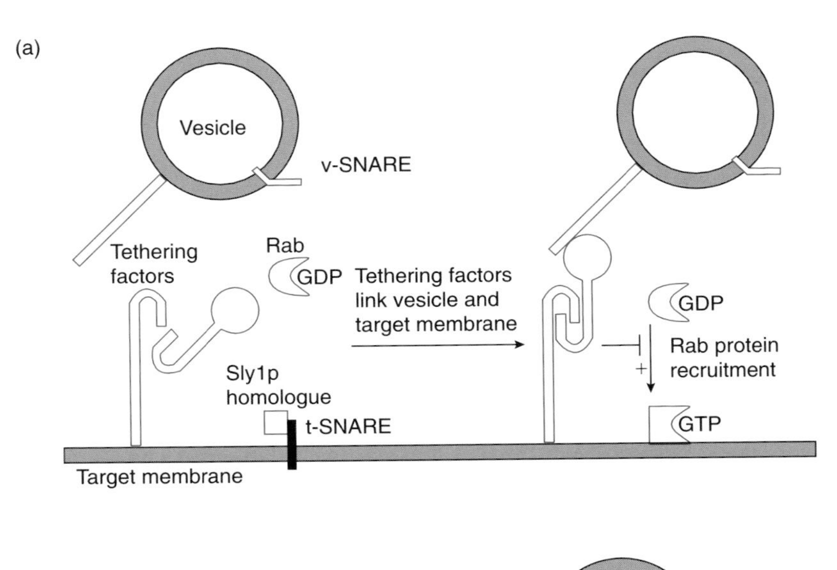

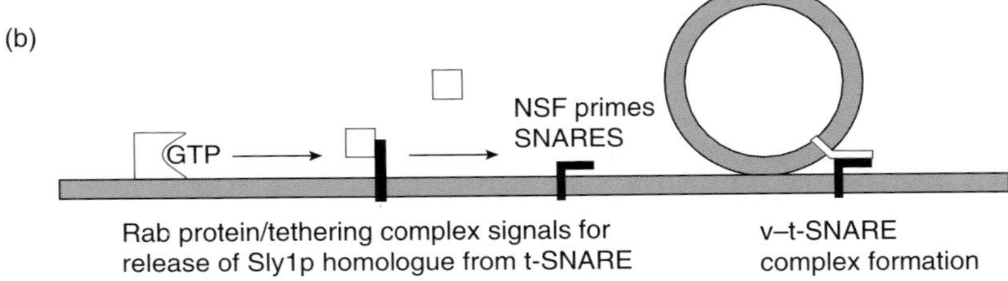

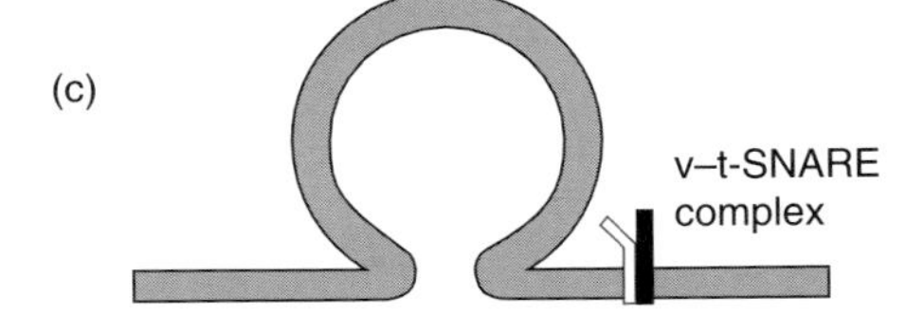

Figure 2. Vesicle tethering and docking
(a) Tethering. As the vesicle approaches its target, membrane-tethering factors form a bridge between the two, in a process dependent on a specific Rab protein. (b) Docking. Activation of SNAREs [soluble *N*-ethylmaleimide-sensitive fusion protein (NSF) attachment protein receptor] requires priming by NSF, subsequently vesicle (v)-SNARE–target-membrane (t)-SNARE complex formation occurs, thus positioning the vesicle ready for membrane fusion. (c) Fusion. Formation of the v-SNARE–t-SNARE complex brings the vesicle and target membranes sufficiently close that bilayer fusion can occur.

[26]. Like p115/Uso1p, both Giantin and GM130 are coiled-coil proteins predicted to have extended structures stretching up to 150 nm from the surface of the membrane. If the structure predictions are correct then these could bridge large distances between incoming vesicles and their target membranes, and initiate the recognition process that leads to specific docking [4]. Not all

5

Functions and origins of the chloroplast protein-import machinery

Danny J. Schnell[1]

Department of Biological Sciences, Rutgers, The State University of New Jersey, 101 Warren Street, Newark, NJ 07102, U.S.A.

Introduction

Chloroplasts represent one of a functionally and structurally diverse group of organelles, the plastids, that are the hallmarks of plant cells. Chloroplasts are the best known of the plastid types due to their central role in photosynthesis and their essential function in amino acid and lipid metabolism in the green tissues of plants. It is generally accepted that plastids originated when a nucleated cell engulfed a photosynthetic prokaryote similar to a modern cyanobacterium. The energy-generating systems of the prokaryote provided important benefits for the host cell, and the resident was retained and assimilated by the process of endosymbiosis. One of the major results of endosymbiosis was the transfer of the majority of endosymbiont genes to the nuclear genome of the host cell. Although the reason for this mass transfer of genetic material is debatable, a major consequence at the cellular level was the development of a compensatory protein-import system to target nuclear-encoded chloroplast proteins back to the organelle. In modern vascular plants, it is estimated that 90% of chloroplast proteins are nuclear-encoded. These proteins are translated on free polysomes in the cytoplasm and are imported into the organelle after their synthesis is complete.

[1]*e-mail: schnell@andromeda.rutgers.edu*

The chloroplast is a complex organelle with three membrane systems: a double membrane envelope that regulates communication between the organelle and the cytoplasm, and an internal thylakoid membrane that harbours the light-harvesting and electron-transport systems of photosynthesis. These three membranes enclose three soluble compartments. Targeting to each subcompartment is directed by targeting signals that are intrinsic to the primary structure of the nuclear-encoded protein [1] (Figure 1). Proteins that must cross more than one membrane *en route* to their final destination, such as thylakoid-membrane and luminal proteins, contain multiple targeting signals that are decoded in sequence as the protein is translocated across successive membranes.

The nature of the targeting signals that direct proteins across the envelope and to the thylakoid membrane have been defined, and a number of components of the translocation systems that operate at these membranes have been identified. Analysis of the structures of the targeting components suggest that these systems have been adapted from transport systems of the original endosymbiont that are comparable with those operating in modern prokaryotes. This chapter will summarize the characteristics of the envelope and thylakoid translocation pathways with a focus on the insights provided by comparisons with prokaryotic systems. Due to space limitations, only the most

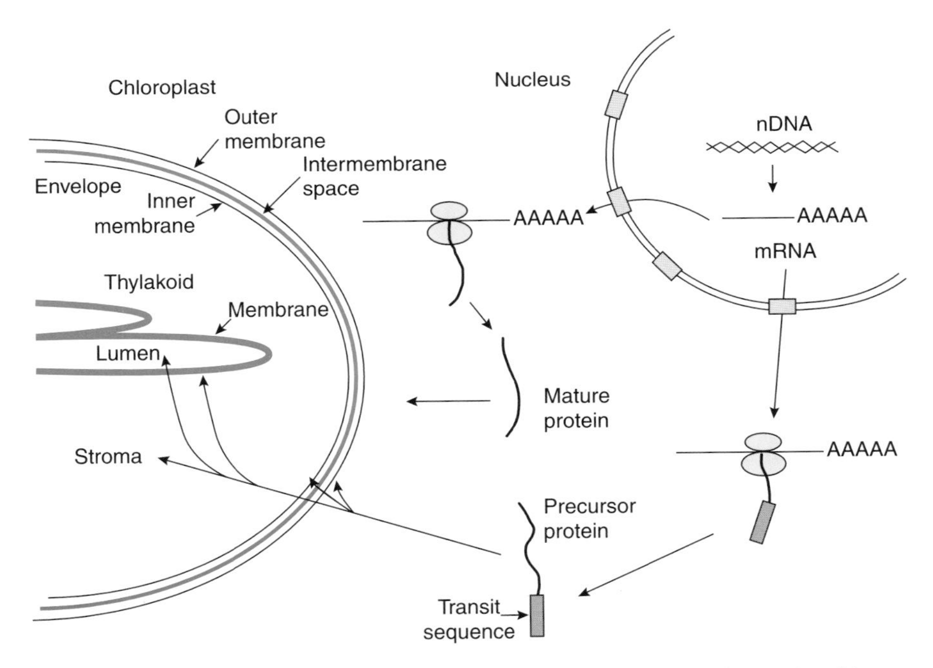

Figure 1. The major targeting pathways of nuclear-encoded proteins to the chloroplast
nDNA, nuclear DNA.

recent or directly relevant references will be cited. In many cases, these will be excellent reviews that provide a detailed analysis of the subject. The reader is asked to refer to these articles for additional references.

Translocation at the chloroplast envelope

Translocation across the envelope

The majority of nuclear-encoded chloroplast proteins are synthesized in the cytoplasm as preproteins containing an N-terminal extension called a transit sequence (Figure 2). Transit sequences are found on all proteins destined for the internal compartments of the organelle, including inner-envelope-membrane, stromal and thylakoid proteins [1]. All transit sequences contain one common primary targeting signal, the stromal-targeting domain (STD), which mediates translocation across the double membrane of the envelope. The transit sequences of stromal proteins are the simplest in structure, consisting solely of an STD (Figure 2). Although the STDs of different preproteins are functionally interchangeable, their structures vary

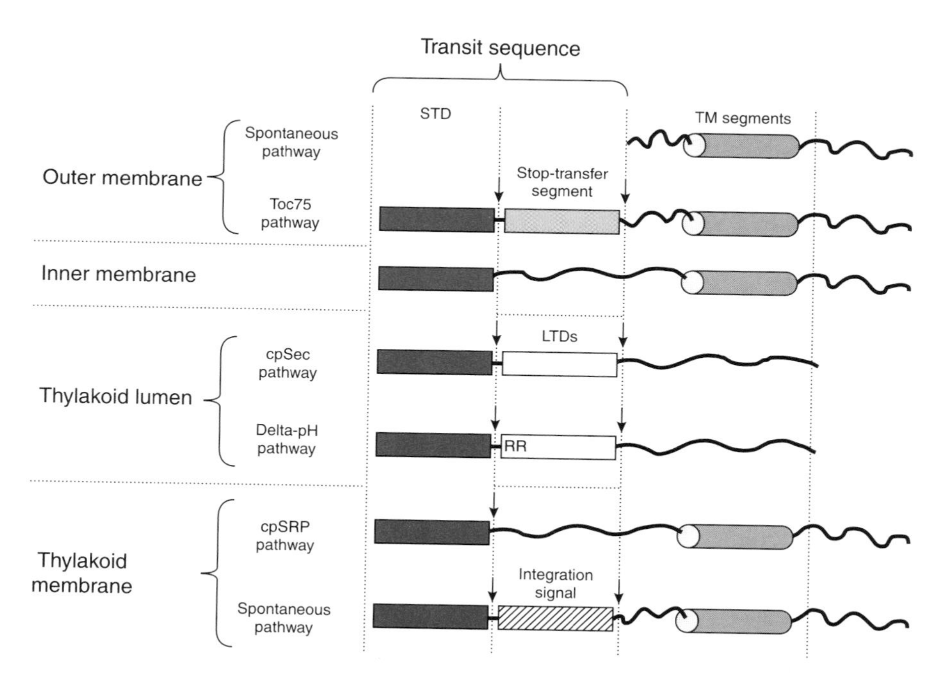

Figure 2. Diagrammatic representation of the major targeting signals that operate in translocation at the three membranes of chloroplasts
Cleavable targeting signals of transit sequences are represented by shaded blocks. These include stromal-targeting domains (STD), thylakoid luminal targeting domains (LTDs), and thylakoid-membrane integration signals. The LTD of the delta-pH pathway is distinguished by the presence of a twin arginine motif (RR). The sites of proteolytic processing are indicated by vertical arrows. Transmembrane (TM) segments that function in targeting are represented by shaded cylinders. Refer to the text for a detailed description of each targeting signal.

considerably. The length of these sequences ranges from 25 to 75 amino acids, and they share no apparent conservation in primary structure. However, STDs do possess several general characteristics, including a prevalence of hydroxylated amino acids, a deficiency in acidic amino acids and an overall basic charge.

It appears that all preproteins that contain a transit sequence are initially translocated across the double membrane of the chloroplast envelope through common, general import machinery. Components of the general import machinery at the outer membrane are given the designation Toc (translocon at the outer-envelope membrane of chloroplasts), whereas the inner-membrane components are given the designation Tic (translocon at the inner-envelope membrane of chloroplasts) [2] (Figure 3a). In this nomenclature, the acronyms are followed by the molecular mass of the protein in kDa. Table 1 presents a list of proteins that have been implicated in envelope translocation; this section will highlight only those for which there is considerable evidence for a direct involvement in import (see [1] for additional references).

Translocation at the outer membrane is an energy-dependent process, requiring both GTP and ATP hydrolysis at the chloroplast surface or within the intermembrane space [1] (Figure 3a). A set of integral membrane proteins, the Toc complex, mediates preprotein recognition and translocation. Two Toc

Table 1. Components of the translocation apparatus at the chloroplast envelope

Component	Proposed function
Outer membrane	
Toc components	
Toc159	Preprotein recognition; protein conductance
Toc75	Preprotein recognition; protein conductance
Toc40	Unknown
Toc34	Regulation of preprotein recognition
Chaperones	
Com70	Prevention of preprotein folding at the chloroplast surface
Hsp70-IAP	Prevention of preprotein folding in the intermembrane space
Inner membrane	
Tic components	
Tic20	Protein conductance
Tic22	Preprotein recognition at the inner-membrane translocon
Tic55	Unkown
Tic110	Docking site for stromal chaperones
Stromal chaperones	
ClpC	Driving force for inner-membrane translocation; protein folding
Cpn60	Protein folding

GTPases, Toc159 and Toc34, appear to regulate recognition and translocation at the outer membrane through cycles of GTP binding and/or hydrolysis [3]. Toc159 appears to be the major transit sequence receptor at the Toc complex, and the GTPase activity of Toc34 is required for insertion of preproteins

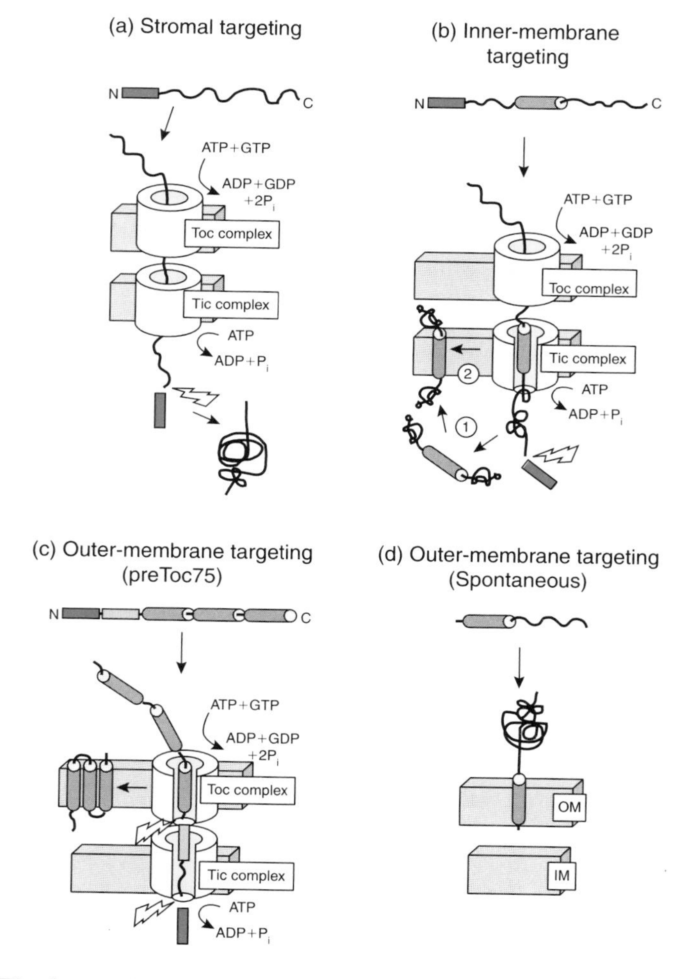

Figure 3. The four major targeting pathways into and across the chloroplast envelope

The targeting signals for each pathway are represented as blocks or cylinders corresponding to those given in Figure 2. The lightening bolt indicates transit-sequence processing. (a) Translocation into the stroma occurs through the co-ordinated action of the Toc and Tic complexes of the general import machinery in the outer- and inner-envelope membranes, respectively. (b) Inner-envelope membrane proteins initiate translocation through the general import machinery. Integration into the inner membrane is proposed to occur through a stromal intermediate (pathway 1) or by lateral insertion into the membrane through the Tic complex (pathway 2). (c) PreToc75 begins translocation through the general import machinery, but a stop-transfer signal triggers its integration into the outer membrane. (d) Proteins using the spontaneous pathway integrate directly into the outer-envelope membrane (OM) without the need for proteinaceous receptors or an exogenous energy source. IM, inner-envelope membrane.

across the outer membrane. A third component, Toc75, is a porin-like ion channel that forms an essential part of the outer-membrane protein conducting channel [4]. In addition to the translocon components, two molecular chaperones of the Hsp70 family, Hsp70-IAP and Com70, associate with preproteins at the early stages of translocation [1]. The Hsp70-IAP (Hsp70-import intermediate-associated protein) is tightly associated with the inner face of the outer membrane, whereas Com70 is peripherally bound to the outer face of the outer membrane. Chaperone binding is likely to stabilize preproteins as they insert across the membrane, and may aid in their vectorial transport through cycles of ATP hydrolysis.

Upon inserting across the outer membrane, preproteins associate with Tic components at the inner-envelope membrane and translocation into the stroma proceeds at the expense of stromal ATP (Figure 3a). Tic20 and Tic22 covalently cross-link to preproteins during envelope translocation and are proposed to participate directly in the translocation process [5]. A third integral inner-membrane protein, Tic110, serves as a docking site for the stromal chaperones, Hsp93 (ClpC) and the GroEL homologue Cpn60 [6,7]. The role of the chaperones is under investigation, but they are likely to participate in translocation and folding of newly imported proteins.

Translocation of preproteins across the outer- and inner-envelope membranes is simultaneous. Components of the Toc and Tic complexes associate directly to form functional contact sites that provide a direct path for preproteins from the cytoplasm to the stroma (Figure 3a). It is likely that preprotein binding triggers association of the Toc and Tic components although a bound preprotein is not required for the association [6]. During or shortly after translocation, the transit sequence of stromal proteins is removed by a specific stromal processing protease to yield the mature proteins [8].

Integration into the envelope membranes

Integral inner-envelope membrane proteins contain transit sequences that function as STDs to initiate translocation across the outer- and inner-envelope membranes (Figure 2). Integration into the inner membrane is directed by one or more transmembrane segments of the protein (Figure 2). Two models have been proposed for the mechanism of integration. The triose phosphate/phosphate translocator and a 37 kDa protein of unknown function are proposed to integrate into the membrane by a membrane stop-transfer mechanism [9]. In this model, the hydrophobic region of the transmembrane segment triggers lateral insertion of the proteins into the inner membrane prior to complete envelope translocation (Figure 3b). The second model is based on the observation that chimaeric proteins containing the transmembrane segment of Tic110 and a soluble passenger protein integrate into the inner membrane via a soluble stromal intermediate [10]. These data suggest that Tic110 may fully translocate across the envelope and reinsert into the inner-envelope membrane from the stromal compartment (Figure 3b).

Most outer-envelope membrane proteins appear to insert directly into the outer membrane by a pathway independent of transit sequences (for review see [1]; Figure 3d). These proteins are synthesized in the cytoplasm as mature polypeptides, and their insertion into the outer membrane is directed by sequences adjacent to and including their transmembrane anchors (Figure 2). One outer-envelope membrane protein, Toc75, is an exception to this generalization. Toc75 contains a presequence that is essential for targeting to the outer membrane (Figure 2). The N-terminal domain of the presequence functions as an STD when transferred to a passenger protein. However, the C-terminal domain of the presequence acts as a stop-transfer domain that prevents complete translocation into the stroma and triggers integration of Toc75 into the outer membrane [11] (Figure 3c).

Translocation at the thylakoid membrane

Thylakoid luminal proteins

Thylakoid luminal proteins possess bipartite transit sequences that contain two independent targeting signals [12]. The N-terminal domain of the transit sequence functions as a typical STD, whereas the C-terminal domain functions as a thylakoid luminal targeting domain (LTD; Figure 2). The STD is cleaved by the soluble chloroplast-processing enzyme after envelope translocation, generating a soluble stromal intermediate that retains the LTD. LTDs share the features of a short basic N-terminal region, a hydrophobic core of 12–18 amino acids and a polar C-terminal region. These three regions are analogous to the domain structures of bacterial signal peptides, indicating that the pathways for targeting of these proteins are homologous to protein-export pathways that operate across bacterial plasma membranes. For this reason, the LTDs often are referred to simply as signal peptides. In fact, LTDs are cleaved by a thylakoid luminal protease that closely resembles the periplasmic leader peptidase that cleaves bacterial signal peptides.

The LTDs of thylakoid luminal proteins can be subdivided into two distinct classes based on distinguishing structural features. The LTDs of preplastocyanin and preOE33 represent the first class. These LTDs most closely resemble signal peptides that direct translocation on the general export or Sec pathway in bacteria. The second class of LTDs is found within the transit sequences of luminal proteins such as preOE23 and preOE16. Although these LTDs contain a similar domain structure to signal peptides, they are distinguished by the presence of a twin arginine motif (RR) immediately adjacent to the N-terminal border of the hydrophobic core of the targeting signal [13] (Figure 2).

The two classes of LTDs for thylakoid luminal preproteins correspond to two pathways for translocation with distinct components and energy requirements (Figure 4). Preproteins containing LTDs that most closely resemble classical signal peptides (e.g. preplastocyanin and preOE33) strictly require

Table 2. Components of the translocation apparatus of the thylakoid

Component	Proposed function
Thylakoid luminal targeting	
cpSec pathway	
cpSecA	Stromal-targeting factor; homologous to bacterial SecA
cpSecY	Preprotein conductance; homologous to SecE component of bacterial SecYEG translocon
ΔpH pathway	
Hcf106	Preprotein recognition
Thylakoid membrane targeting	
cpSRP pathway	
cpSRP54	Subunit of stromal-targeting factor; homologous to bacterial Ffh and mammalian SRP54
cpSRP43	Subunit of stromal-targeting factor
cpFtsY	cpSRP receptor; homologous to bacterial FtsY

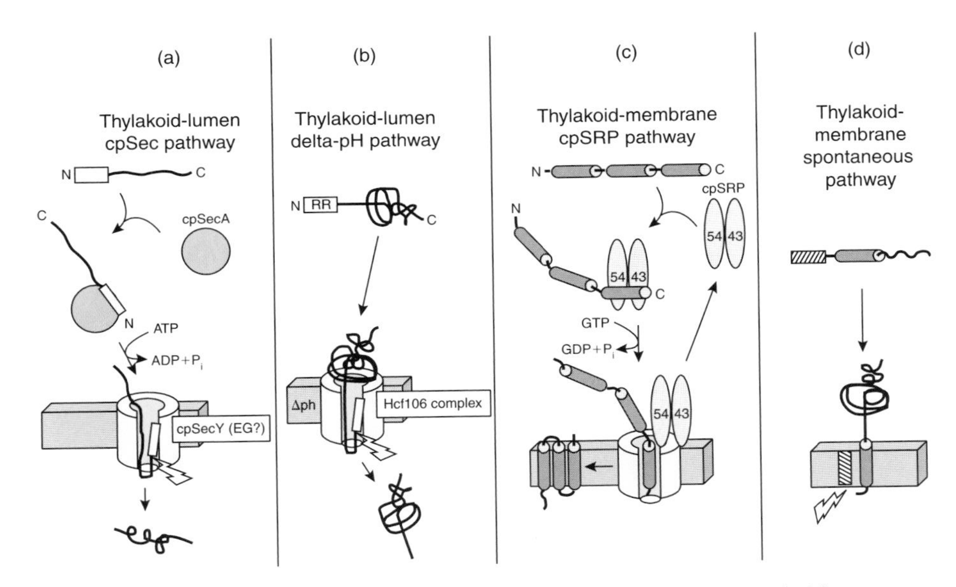

Figure 4. The four major targeting pathways into and across the thylakoid membrane

The targeting signals for each pathway are represented as blocks or cylinders corresponding to those given in Figure 2. (a) The cpSec pathway is a homologue of the bacterial preprotein translocase. Transport requires ATP hydrolysis. (b) The delta-pH pathway requires a transthylakoidal pH gradient and can transport folded proteins. (c) The cpSRP pathway requires GTP hydrolysis and involves cpSRP, which has similarity with the bacterial and mammalian signal recognition particles. (d) The spontaneous pathway does not require an obvious energy source or proteinaceous receptors at the thylakoid membrane.

ATP for translocation (see [12] for review). At least two chloroplast proteins have been identified that participate in translocation, cpSecA [14] and cpSecY [15] (Table 2). These proteins are homologous to the SecA and SecY components of the Sec machinery that participate in the general protein-export pathway in prokaryotes. As a result, this pathway is designated the cpSec pathway (Figure 4a). In bacteria, SecA serves as a soluble signal-sequence receptor that delivers preproteins to a membrane-bound translocation system, the SecYEG complex.

The second pathway for luminal targeting is designated the ΔpH pathway due to the requirement of a transthylakoidal pH gradient for translocation (for a summary see [12]; Figure 4b). This pathway is selective for preproteins containing a twin arginine motif within their LTDs (Figure 2). One other distinguishing feature of this pathway is its ability to translocate proteins that are partially or fully folded while maintaining the critical transmembrane potential of the thylakoid membrane [16]. This is a unique feature of the pathway because membrane translocation generally requires an unfolded substrate. The gene for one component of the ΔpH pathway, *hcf106*, has recently been identified as a maize mutant defective in thylakoid biogenesis [17]. The Hcf106 protein is an integral thylakoid membrane protein with a single transmembrane domain (Table 2). A number of genes encoding proteins with similarity to Hcf106 have been identified in prokaryotes. At least one of the bacterial genes has been implicated in the export of redox proteins with signal peptides containing twin arginine motifs similar to the thylakoid proteins [16]. Remarkably, the redox proteins fold and bind their cofactors prior to export across the plasma membrane. Consequently, it appears that the thylakoidal ΔpH pathway was adapted from a pre-existing pathway distinct from the general Sec pathway.

Thylakoidal membrane proteins

Two types of signal also exist to target nuclear-encoded integral membrane proteins to the thylakoid membrane. The first type of signal is represented by the precursor to the light-harvesting chlorophyll a/b-binding protein (preLHCP). In contrast to other thylakoid proteins, the transit sequence of preLHCP consists only of an STD with no additional information for thylakoid targeting. Targeting to the thylakoid membrane is contained within the mature portion of the polypeptide and is localized to one of its three transmembrane domains (Figure 2). Targeting to the membrane requires GTP and involves a transit complex that forms in the stroma after the transit sequence of LHCP has been cleaved [18] (Figure 4c). Two components of the transit complex have been identified. One protein, cpSRP54 (Table 2), is related to the 54 kDa GTP-binding subunit of the mammalian signal-recognition-particle (SRP) complex and its bacterial counterpart, Ffh. The mammalian SRP and bacterial Ffh direct co-translational targeting of nascent polypeptide–ribosome complexes to the endoplasmic reticulum and bacterial plasma membrane, respectively. The second protein of the targeting complex, cpSRP43 (Table 2), is a unique polypeptide with no homology with any known protein. cpSRP54 and cpSRP43 bind to transmembrane regions of LHCP as a complex and target the protein to the membrane with the aid of an additional unidentified stromal factor and GTP. The bacterial Ffh targeting complex binds to a cognate receptor, FtsY, and delivers the preprotein to the SecYEG complex for translocation across the plasma membrane. A chloroplast homologue of FtsY exists, but it remains to be shown whether it functions in co-operation with cpSRP or whether LHCP integration involves the Sec machinery.

The second type of membrane-targeting signal involves proteins that are synthesized with bipartite transit sequences (e.g. the Cf_0II subunit and photosystem II subunits W and X). The overall structure of the transit sequences are similar to the bipartite structure of luminal proteins, with an N-terminal STD followed in tandem with a signal-peptide-like thylakoid-targeting signal. However, the thylakoid-targeting domain does not function as a signal sequence, and therefore represents a unique integration signal. The targeting mechanism of this class of proteins is unusual in that integration does not require proteinaceous factors or an obvious energy source [1]. As a result, this pathway has been termed the 'spontaneous pathway' (Figure 4d). The mechanism of spontaneous integration appears to rely on direct interactions of the hydrophobic integration signal with the lipid bilayer. In this respect, it resembles the integration of the M13 coat protein into the *Escherichia coli* plasma membrane, and targeting of a number of integral proteins to the outer membranes of chloroplasts and mitochondria.

Perspectives

The picture that emerges from investigations into protein import into chloroplasts is of a set of independent targeting pathways that act in series to direct nuclear-encoded proteins to the proper chloroplast subcompartment. Elements of the translocation systems at both the envelope and thylakoid membranes appear to have been adapted from pre-existing secretion systems present in the original prokaryotic endosymbiont. Each of the systems has evolved to varying degrees depending on the requirements of endosymbiosis.

The cpSec system appears to be remarkably similar to the prokaryotic Sec apparatus. This is not surprising because the topological constraints that govern the translocation of stromal intermediates of luminal proteins (e.g. plastocyanin) are comparable with those for bacterial proteins that follow the general export pathway. The ΔpH pathway appears to have been adapted to translocate proteins that partially or completely fold in the stroma prior to translocation into the thylakoid. These systems present a challenge to the conventional view of translocation in which an unfolded polypeptide is transported linearly as a polyion through a channel that acts much like a regulated ion channel. The ability of the ΔpH pathway to transport folded substrates indicates a remarkable ability of the protein-conducting component to expand and accommodate the diameter of a folded polypeptide while maintaining a tight seal at the membrane. Furthermore, there is no clear precedent for the electrochemical gradient as the sole energy source for translocation. How the pH gradient is coupled to transport remains an intriguing question.

The cpSRP54–43 system appears to have undergone extensive adaptation from the original endosymbiont. The cpSRP54 has lost the 7 S RNA component that is essential for the function of Ffh and the cytoplasmic SRP, but it has gained an additional protein subunit, cpSRP43, which is unique to the thylakoid system. These changes have been attributed to the fact that the cytoplasmic and bacterial SRPs function co-translationally, whereas LHCP targeting to the thylakoid is strictly post-translational. Remarkably, cpSRP54, but not cpSRP43, recently has been shown to participate in co-translational targeting of plastid-encoded thylakoid proteins [19]. Thus cpSRP54 may have retained a function similar to Ffh, but has been adapted to function in the LHCP targeting pathway by the addition of a unique 43 kDa subunit.

Perhaps the most dramatic evolutionary adaptation has occurred at the envelope translocation machinery. A protein-import system is not known to function in Gram-negative bacteria. As a result, the chloroplast-envelope translocation machinery would have had to evolve *de novo* or be adapted from a system with a distantly related function. The discovery of cyanobacterial homologues of Toc75, Tic22 and Tic20 suggests that the latter of the two possibilities is the most likely [20]. This raises the possibility that a cyanobacterial export system might have evolved to form a protein-conducting channel that operates in the opposite direction in chloroplasts. Homologues of Toc159 and Toc34 are not present in prokaryotes or other systems, suggesting that these components evolved from the nuclear genome to provide a system of preprotein recognition and control vectorial translocation.

Addressing the remaining issues in chloroplast import should be greatly facilitated by the existence of related bacterial systems. The prokaryotic systems provide the opportunity for comparative studies, and the ability to do genetic complementation studies to investigate the divergence of the systems over evolution. These approaches will complement and extend the elegant biochemical and genetic studies in plants that have laid the groundwork for our understanding of the chloroplast import process.

Summary

- *The vast majority of chloroplast proteins are nuclear-encoded and are imported into the organelle after synthesis in the cytoplasm.*
- *Targeting to chloroplasts is mediated by a variety of intrinsic targeting signals that direct the preprotein to its proper organelle subcompartment.*

- *Translocation at the envelope membrane is directed by the interactions of an N-terminal transit sequence on the preprotein and a general import machinery composed of the outer-membrane Toc machinery and the inner-membrane Tic machinery. The Toc and Tic components interact to bypass the intermembrane space and provide direct transport of preproteins from the cytoplasm to the stroma.*
- *There are at least four targeting pathways to the thylakoid membrane, the cpSec pathway, the ΔpH pathway, the cpSRP pathway and the spontaneous pathway. These pathways require distinct intrinsic targeting signals, and apparently evolved to accommodate the translocation of classes of proteins with particular characteristics.*
- *Proteins similar to some components of the envelope and thylakoid translocation pathways are found in bacterial systems. However, a number of components do not have bacterial counterparts and are unique to the chloroplast pathways. It therefore appears that the chloroplast translocation systems have evolved from membrane-transport systems that were present in the original endosymbiont by incorporating proteins necessary to adapt to the constraints of endosymbiosis.*

References

1. Keegstra, K. & Cline, K. (1999) Protein import and routing systems of chloroplasts. *Plant Cell* **11**, 557–570
2. Schnell, D.J., Blobel, G., Keegstra, K., Kessler, F., Ko, K. & Soll, J. (1997) A nomenclature for the protein import components of the chloroplast envelope. *Trends Cell Biol.* **9**, 222–227
3. Chen, K., Chen, X. & Schnell, D.J. (2000) Initial binding of preproteins involving the Toc159 receptor can be bypassed during protein import into chloroplasts. Plant Physiol. **122**, 813–822
4. Hinnah, S.C., Hill, K., Wagner, R., Schlicher, T. & Soll, J. (1997) Reconstitution of a chloroplast protein import channel. *EMBO J.* **16**, 7351–7360
5. Kouranov, A., Chen, X., Fuks, B. & Schnell, D.J. (1998) Tic20 and Tic22 are new components of the protein import apparatus at the chloroplast inner envelope membrane. *J. Cell Biol.* **143**, 991–1002
6. Nielsen, E., Akita, M., Davila-Aponte, J. & Keegstra, K. (1997) Stable association of chloroplastic precursors with protein-translocation complexes that contain proteins from both envelope membranes and a stromal Hsp100 molecular chaperone. *EMBO J.* **16**, 935–946
7. Kessler, F. & Blobel, G. (1996) Interaction of the protein import and folding machineries in the chloroplast. *Proc. Natl. Acad. Sci. U.S.A.* **93**, 7684–7689
8. Richter, S. & Lamppa, G.K. (1998) A chloroplast processing enzyme functions as the general stromal processing peptidase. *Proc. Natl. Acad. Sci. U.S.A.* **95**, 7463–7468
9. Brink, S., Fischer, K., Klosgen, R.-B. & Flugge, U.-I. (1995) Sorting of nuclear-encoded chloroplast membrane proteins to the envelope and the thylakoid membrane. *J. Biol. Chem.* **270**, 20808–20815
10. Lubeck, J., Heins, L. & Soll, J. (1997) A nuclear-encoded chloroplastic inner envelope membrane protein uses a soluble sorting intermediate upon import into the organelle. *J. Cell Biol.* **137**, 1279–1286
11. Tranel, P.J. & Keegstra, K. (1996) A novel, bipartite transit peptide targets OEP75 to the outer membrane of the chloroplastic envelope. *Plant Cell.* **8**, 2093–2104

12. Schnell, D.J. (1998) Protein targeting to the thylakoid membrane. *Annu. Rev. Plant Physiol. Plant Mol. Biol.* **49**, 97–126
13. Chaddock, A.M., Mant, A., Karnauchov, I., Brink, S., Herrmann, R.G., Klosgen, R.B. & Robinson, C. (1995) A new type of signal peptide: central role of a twin-arginine motif in transfer signals for the delta-pH-dependent thylakoidal protein translocase. *EMBO J.* **12**, 2715–2722
14. Yuan, J., Henry, R., McCaffery, M. & Cline, K. (1994) SecA homolog in protein transport within chloroplasts: evidence for endosymbiont-derived sorting. *Science* **266**, 796–798
15. Roy, L.M. & Barkan, A. (1998) A SecY homologue is required for the elaboration of the chloroplast thylakoid membrane and for normal chloroplast gene expression. *J. Cell Biol.* **141**, 385–395
16. Stephens, C. (1998) Protein secretion: getting folded proteins across membranes. *Curr. Biol.* **13**, 578–581
17. Settles, A.M., Yonetani, A., Baron, A., Bush, D.R., Cline, K. & Martienssen, R. (1997) Sec-independent protein translocation by the maize Hcf106 protein. *Science*. **278**, 1467–1470
18. Schuenemann, D., Gupta, S., Persello-Cartieaux, F., Klimyuk, V.I., Jones, J.D., Nussaume, L. & Hoffman, N.E. (1998) A novel signal recognition particle targets light-harvesting proteins to the thylakoid membranes. *Proc. Natl. Acad. Sci. U.S.A.* **95**, 10312–10316
19. Nillson, R., Brunner, J., Hoffman, N.E. & van Wijk, K.J. (1999) Interactions of ribosome nascent chain complexes of the chloroplast-encoded D1 thylakoid membrane protein with cpSRP54. *EMBO J.* **18**, 733–742
20. Reumann, S. & Keegstra, K. (1999) The endosymbiotic origin of the protein import machinery of the chloroplastic envelope membranes. *Trends Plant Sci.* **4**, 302–307

6

Mechanisms of mitochondrial protein import

Donna M. Gordon*, Andrew Dancis† and Debkumar Pain*[1]

**Department of Physiology, University of Pennsylvania School of Medicine, D403 Richards Building, 3700 Hamilton Walk, Philadelphia, PA 19104-6085, U.S.A., and †Department of Medicine, University of Pennsylvania School of Medicine, Philadelphia, PA 19104-6100, U.S.A.*

Introduction

Mitochondria are the organelles that mediate respiration and ATP synthesis in eukaryotic cells. They also participate in numerous indispensable metabolic pathways (e.g. synthesis of haem, nucleotides, lipids and amino acids) and they mediate intracellular homoeostasis of inorganic ions. Mitochondrial function is, therefore, essential for the viability of eukaryotic cells.

New mitochondria are formed by the growth and division of pre-existing organelles. Although mitochondria contain their own DNA and complete systems for its replication, transcription and translation, they synthesize only a few proteins (13 in humans). All other mitochondrial proteins are nuclear-encoded and are synthesized on cytoplasmic ribosomes. These proteins must be transported from the cytosol to the correct mitochondrial subcompartment. There are four such subcompartments: outer membrane (OM), intermembrane space (IMS), inner membrane (IM) and matrix. Protein import into mitochondria is therefore a fundamental mechanism of mitochondrial biogenesis. Much of the information on mitochondrial protein import has come from studies of the yeast, *Saccharomyces cerevisiae*, and the mold, *Neurospora crassa*. It is like-

[1]*To whom correspondence should be addressed (e-mail: pain@mail.med.upenn.edu).*

ly, however, that similar mechanisms function in human mitochondria, since human mitochondrial precursor proteins can be efficiently imported into yeast mitochondria and human homologues of yeast mitochondrial protein translocases have been identified [1,2].

Mitochondrial targeting and sorting signals

Most matrix proteins are synthesized as preproteins with an N-terminal extension known as the signal sequence, targeting sequence or presequence (Figures 1 and 2). In general, signal sequences are 15–30 amino acids long and

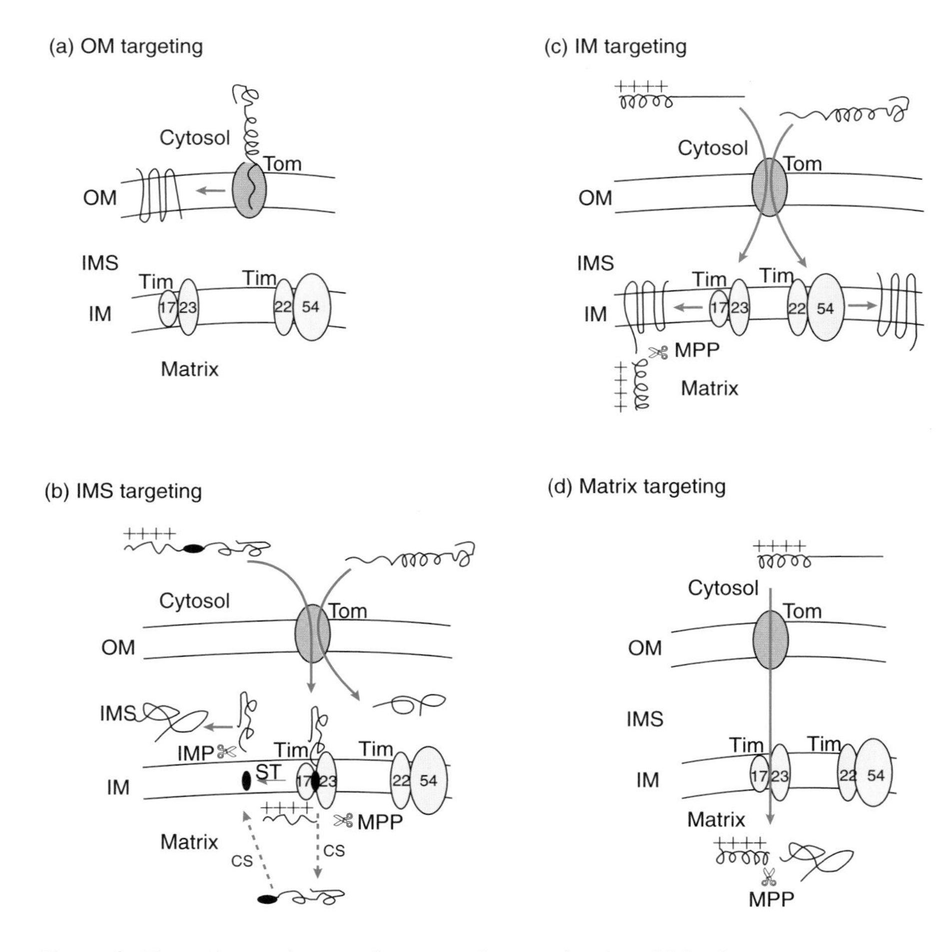

Figure 1. Targeting pathways of preproteins to mitochondrial subcompartments
Targeting pathways for nuclear-encoded mitochondrial preproteins with or without a cleavable signal sequence are shown. The cleavable basic signal sequence located at the N-terminus of a preprotein is indicated by ++++. Tom, translocase of OM; Tim, translocase of IM; IMP, IM proteases; MPP, matrix-processing peptidase. Two different Tim complexes are indicated by Tim17-23 and Tim22-54. For (b) IMS targeting, CS and ST indicate conservative-sorting and stop-transfer pathways, respectively.

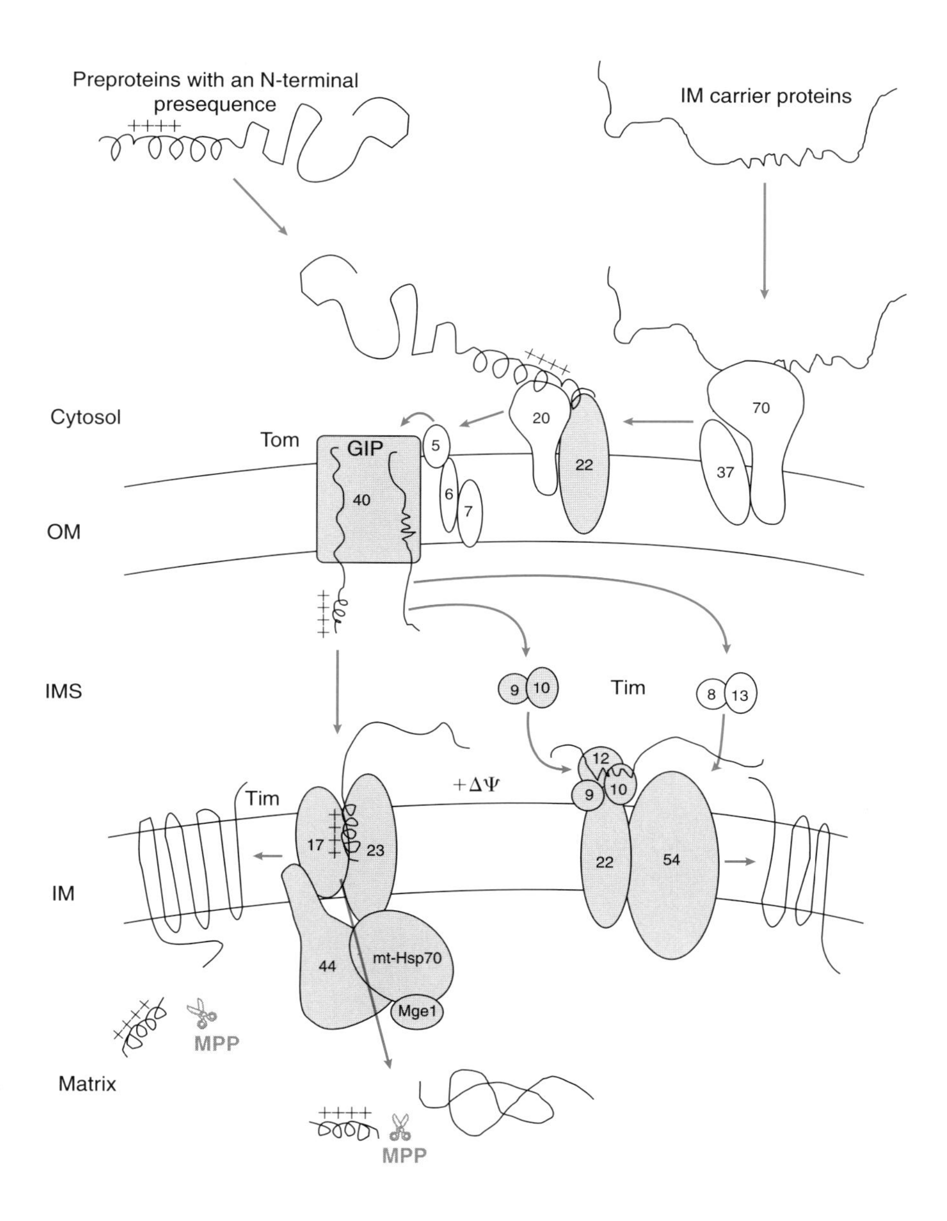

Figure 2. Schematic model of the mitochondrial protein-import machinery
Protein translocation across the OM is mediated by a general Tom complex. Following translocation across the OM, preproteins with cleavable signal sequences are recognized by the general Tim17-23 machinery for either insertion into the IM or continued translocation into the matrix (left-hand side). On the other hand, carrier proteins with internal targeting information are recognized by the specialized Tim22-54 machinery for insertion into the IM (right-hand side). The essential components are shaded. $\Delta\Psi$ denotes the membrane potential across the IM; Tom, translocase of OM; Tim, translocase of IM; MPP, matrix-processing peptidase; GIP, general insertion pore; mt-Hsp70, mitochondrial Hsp70.

are rich in positively charged (usually Arg) and hydroxylated (Ser and Thr) residues; they usually lack acidic amino acids. Comparison of the primary structures of these signal sequences reveals no obvious homology or motif that might be responsible for the targeting function. These signal sequences, however, can adopt an amphiphilic α-helix that might be important for their recognition by the mitochondrial protein-import machinery. The signal sequence is both necessary and often sufficient for directing a protein into mitochondria [1,2]. Upon import, the signal sequence is removed by matrix-localized signal peptidase(s) (see below).

Proteins like cytochrome b_2 and cytochrome c_1 are synthesized with unusually long presequences (60–80 amino acids) that contain two domains: the N-terminal hydrophilic domain, which resembles a matrix-targeting signal, and the C-terminal hydrophobic domain responsible for sorting to the IMS. Cleavage of this long presequence occurs in two steps, the first being catalysed by the matrix-processing peptidase (MPP), the second by peptidase(s) located at the outer face of the IM (Imp1p and Imp2p). A particularly challenging question concerns how mitochondria decode the IMS sorting information [1]. According to the 'conservative sorting' model, cytochrome b_2 and cytochrome c_1 are first imported completely into the matrix and then re-exported across the IM to the IMS by a process that resembles protein secretion in bacteria. Alternatively, according to the 'stop-transfer' model, the hydrophobic domain of the presequence arrests translocation through the IM, leaving the mature part of the precursor protein in the IMS without ever crossing the IM (Figure 1b).

On the other hand, many preproteins lack cleavable signal sequences, and instead carry their targeting information in poorly defined internal sequences. This class of proteins includes all OM, some IMS and some IM proteins. Most of these preproteins also contain a hydrophobic sorting signal that interrupts translocation across the OM or the IM, and thereby helps to direct the protein to its final destination in the OM, the IMS or the IM (Figures 1a–c).

Post-translational import and cytosolic chaperones

Mitochondrial precursor proteins can be imported post-translationally [1,2]. If a precursor protein is synthesized *in vitro* in a cell-free translation system (e.g. rabbit reticulocyte lysate), the ribosomes can then be sedimented, leaving the nascent preprotein in the post-ribosomal supernatant. When isolated mitochondria are added to this supernatant, the precursor protein is imported into mitochondria. Pulse–chase experiments *in vivo* also suggest a post-translational mechanism of import; preproteins can be detected in the cytosol before they are imported into the mitochondria. However, *in vivo* it is possible that some proteins are imported in a co-translational manner. For example, when yeast cells are treated with cycloheximide to block the cytosolic protein synthesis, cytoplasmic ribosomes remain attached to mitochondria. The mRNAs associated with these ribosomes are enriched in mRNAs of mitochondrial proteins.

A prerequisite for post-translational import is that preproteins must be maintained in a translocation-competent conformation. In some cases, this is achieved through interactions with cytosolic chaperones, such as members of the Hsp70 family and mitochondrial-import-stimulating factor (MSF). The latter is a heterodimer of 30- and 32-kDa subunits and belongs to the family of 14-3-3 proteins. Although both Hsp70 and MSF are ATPases, they differ in substrate specificity: while Hsp70 can interact with proteins destined for mitochondria as well as other organelles, MSF preferentially interacts with mitochondrial proteins. Mitochondrial preproteins interacting with Hsp70 or MSF are initially transferred to two different receptor subcomplexes that subsequently present the preproteins to the OM channel (see below). Cytosolic chaperones, however, are not required for targeting specificity because chemically pure preproteins can be imported in the absence of any added cytosolic factors [1,2].

Import machinery of mitochondrial membranes

The import of preproteins into mitochondria is mediated by a general translocase in the OM (the Tom complex), which co-operates with two distinct translocases (Tim17-23 and Tim22-54 complexes) in the IM. The number by which a Tom or a Tim protein is designated indicates the molecular mass of that component in kDa. Following translocation across the OM, matrix proteins utilize the Tim17-23 complex whereas the IM carrier proteins utilize the Tim22-54 complex. The Tim17-23 pathway has been characterized extensively over the years whereas the Tim22-54 pathway has been described very recently (Figure 2).

Identification of Tom and Tim components

Both biochemical and genetic approaches have been used to identify Tom and Tim components [1–3]. Examples of the former include: (i) use of specific antibodies that bind to the translocase(s), thereby blocking import, (ii) cross-linking of precursors arrested during translocation and (iii) immunoprecipitation of protein complexes from detergent-solubilized membranes using antibodies against channel components. Genetic methods have included: (i) mislocalization of the cytosolic *URA3* gene product to the mitochondrial matrix by attaching a mitochondrial signal sequence, then selection for mutants with *URA3* activity due to retention of the chimaeric gene product in the cytosol, (ii) selection of temperature-sensitive yeast mutants based on accumulation of the precursor forms of imported proteins, (iii) screens for extragenic or high-copy suppressors of channel-protein mutants that have already been identified and (iv) identification of candidate proteins based on their sequence homology to other proteins.

The Tom complex

The signal sequence of an OM protein initiates translocation through the Tom machinery but then abrogates further translocation so that the protein exits laterally into the plane of the OM (Figure 1a). Proteins destined for other subcompartments are fully translocated across the OM (Figures 1b–1d). The Tom machinery consists of at least eight OM proteins: four receptor subunits (Tom20 [4], Tom22 [5], Tom37 [6] and Tom70 [7]), three small proteins (Tom5 [8], Tom6 [9] and Tom7 [10]) and a main component of the general insertion pore (GIP; Tom40 [1,11,12]). When mitochondria are solubilized with non-ionic detergents, two receptor subcomplexes can be distinguished: Tom22–Tom20 and Tom70–Tom37. These two subcomplexes interact with each other via motifs of 34 amino acids termed the 'tetratricopeptide repeat'. Tetratricopeptide repeat motifs present in Tom70, Tom37 and Tom20 are likely to participate in dynamic protein–protein interactions. The receptor subcomplexes capture preproteins from the cytosol in at least two different ways (Figure 2). The Tom22–Tom20 complex recognizes precursor proteins that are normally bound to cytosolic Hsp70. These preproteins usually have N-terminal cleavable signal sequences. While the negatively charged Tom22 recognizes the positively charged surface of the amphipathic presequence via an electrostatic interaction, Tom20 binds to the non-polar face of the presequence. In a second mechanism, the Tom70–Tom37 complex recognizes precursor proteins which are presented to mitochondria in a complex with MSF. These preproteins usually contain internal targeting information. The precursor bound to the Tom70–Tom37 complex is subsequently transferred to the Tom22–Tom20 complex. The two targeting pathways, however, are not strictly separated, since some preproteins can interact with either Tom22–Tom20 or Tom70–Tom37 complexes. Among the four receptors, Tom22 is the only receptor that is essential for protein import and cell viability. Thus, both *in vitro* and *in vivo*, Tom20, Tom37 and Tom70 can be bypassed without severe functional consequences [1,2].

After associating with the receptors, preproteins move into and across a GIP. This pore is formed by at least four membrane-embedded components: Tom5, Tom6, Tom7 and Tom40. Whereas the first three small Tom proteins are not essential, Tom40 is the main constituent of the GIP and is essential for protein import and cell viability [1,2]. Recently, Pfanner and co-workers have described functional reconstitution of Tom40 [13]. Purified Tom40 forms a hydrophilic cation-selective high-conductance channel with a diameter of approx. 22 Å. This channel specifically binds to and transports mitochondrial signal peptides. The size of the channel diameter indicates that preproteins in their typical folded conformation cannot permeate the channel; they may pass through only in an α-helical or an extended conformation. These results are in good agreement with the numerous observations that a preprotein must be at least partially unfolded prior to import [1].

Tom5 may provide a link between import receptors and the OM import channel [8]. After interaction with the receptors, preproteins are transferred to Tom5. Tom5 has a single OM anchor and a cytosolic segment with a net negative charge. The latter domain perhaps facilitates interactions with positively charged targeting sequences and the subsequent insertion of preproteins into the GIP. Tom6 and Tom7 do not interact directly with the preprotein while it is in transit, but they are likely to modulate the assembly and dissociation of the Tom machinery. Tom6 promotes the association of both the Tom22–Tom20 and Tom70–Tom37 receptor subcomplexes with the import pore, Tom40. On the other hand, Tom7 favours the dissociation of the Tom22–Tom20 complex from Tom40 as well as the dissociation of the Tom22–Tom20 subcomplex itself.

The structural organization of the Tom complex from *N. crassa* mitochondria has recently been described by Neupert and co-workers [14]. Upon reconstitution into liposomes, the purified Tom complex mediates the integration of OM proteins into the lipid bilayer. It is also capable of mediating the import of IMS proteins and of N-terminal signal sequences of matrix-targeted proteins across the bilayer. The Tom complex represents a cation-selective high-conductance channel. Electron microscopy and image analysis of negatively stained Tom complexes demonstrate stain-filled openings with apparent diameters of 20 Å, which may represent import pores and are likely to be formed by Tom40.

At least two hypotheses have been proposed to explain what drives the translocation of preproteins across the OM. The 'acid chain hypothesis' proposes that the translocation across the OM is driven by binding of basic N-terminal presequences to acidic receptors of increasing avidity [15]. A '*cis*-site/*trans*-site' model provides a similar explanation [1]. Binding of preproteins with matrix-targeting signals to the Tom22–Tom20 complex is defined as the *cis*-site binding. During or after translocation across the OM, presequences interact with a *trans*-binding site on the IMS side of the OM. While the molecular nature of the *trans*-binding site remains unclear, it can be distinguished from the *cis*-binding. The *cis* binding is very labile and salt-sensitive, whereas the *trans* binding is much tighter and mostly salt-resistant. The *trans* binding is therefore unlikely to be mediated mainly by ionic interactions. Nevertheless, the presence of two (or more) binding sites with increasing affinity for presequences could at least partially explain transmembrane movement across the OM.

The Tim22-54 machinery for insertion of carrier proteins into the IM

Many polytopic IM proteins (e.g. ADP/ATP carrier) lack a cleavable signal sequence at their N-termini. Instead these proteins contain one or more internal signals. They utilize the general Tom complex for translocation across the OM. At the *trans* side of the OM, these carrier proteins use a specialized Tim22-54 pathway for insertion into the IM (Figure 2). On the other hand, polytopic IM proteins with a cleavable N-terminal signal sequence (e.g. Atm1p, an ATP-binding-cassette transporter) appear to follow the general Tim17-23 matrix pathway (discussed in the next section).

As the N-terminus of the carrier protein becomes exposed to the IMS, it binds to a soluble 70 kDa complex consisting of Tim9 and Tim10 proteins. Such an interaction pulls the carrier protein across the Tom channel. The carrier is subsequently delivered to a 300 kDa membrane-embedded complex which, in addition to a small fraction of Tim9 and Tim10, contains two integral membrane proteins (Tim22 and Tim54) and one peripheral membrane protein (Tim12). This complex catalyses the insertion of the carrier protein into the IM in a membrane-potential ($\Delta\Psi$)-dependent manner. Tim9, Tim10, Tim12, Tim22 and Tim54 are all essential for cell viability. Two other soluble IMS proteins, Tim8 and Tim13, have been identified as additional components that guide other hydrophobic proteins (e.g. Tim23) through the IMS to the IM insertion machinery (Figure 2). Neither Tim8 nor Tim13 is essential for viability. Due to space limitations, original references for Tim proteins described in this section cannot be cited. Instead, readers are referred to a very recent and excellent review that emphasizes in detail the role of these Tim proteins in the biogenesis of mitochondrial IM proteins [3].

The Tim17-23 complex for import into the matrix

The Tim17-23 complex consists of Tim17 [16,17], Tim23 [18,19] and Tim44 [20,21], which are essential elements of the general import pathway for translocation of preproteins across the IM into the matrix. Tim17 and Tim23 are integral membrane proteins and are believed to be the core elements of the IM channel. These two proteins are predicted to span the membrane four times and share sequence similarity in their transmembrane domains. The hydrophilic N-terminal portion of Tim23 is exposed to the IMS. This domain contains a net negative charge and could potentially serve as a sensor that recognizes positively charged signal sequences as they emerge from the *trans* side of the OM channel (Figure 2). Unlike Tim17 and Tim23, Tim44 behaves as a peripheral membrane protein with a large matrix domain; it appears to couple channel function with the energy device that drives the import process. Tim44 recruits mitochondrial Hsp70 (mt-Hsp70) to the site where the preprotein emerges from the Tim channel. mt-Hsp70 is essential for viability and participates not only in protein translocation but also in protein folding in the matrix (described below).

Energy requirements for preprotein translocation into the matrix

The role of membrane potential

Translocation of preproteins into or across the IM is absolutely dependent on a membrane potential ($\Delta\Psi$). The membrane potential exerts its effect early in the series of Tim-mediated processes [1,22]. Only the initial insertion and partial translocation of a presequence across the IM requires $\Delta\Psi$; translocation of the mature part of the preprotein does not require $\Delta\Psi$. It seems likely that the positively charged presequence is electrophoretically 'pulled' through the IM by the $\Delta\Psi$ across this membrane (negative inside).

The role of GTP

The completion of translocation into the matrix is independent of $\Delta\Psi$ but requires a GTP-dependent 'push'. The GTP-dependent push of the polypeptide chain across the IM, in turn, may facilitate unfolding of the C-terminal domains outside the organelle (Figure 3). This push is likely to be mediated by an unknown membrane-bound GTPase on the *cis* (IMS) side of the IM [22].

The role of matrix ATP-dependent interactions

After a sufficient length of the polypeptide chain has penetrated into the matrix due to the continuing GTP-dependent push, mt-Hsp70 comes into play (Figure 3). An efficient unidirectional transmembrane movement of proteins across the IM into the matrix is achieved only through co-ordination of *cis* GTP-dependent processes and *trans* ATP-dependent mt-Hsp70–Tim44 cycles. Both these processes are necessary; neither one can substitute for the other if efficient import into the matrix is to be achieved. It remains to be determined whether the GTP-dependent push is turned off after mt-Hsp70 grabs the incoming polypeptide chain. Alternatively, the process might continue until the import is complete [22].

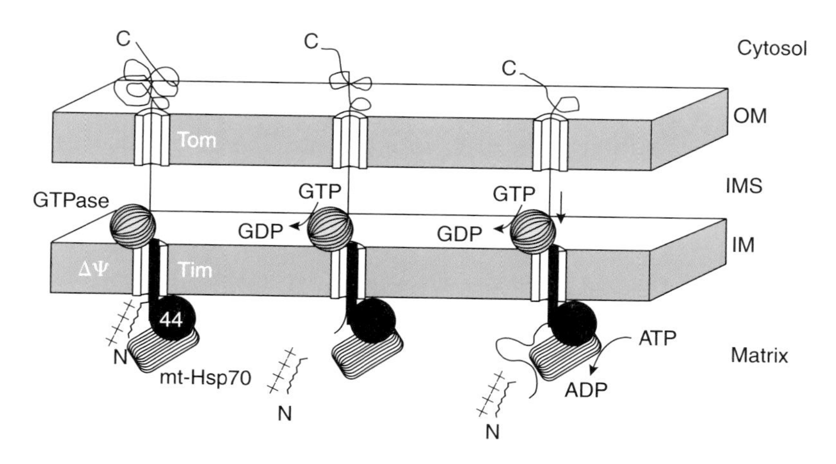

Figure 3. Energy requirements for translocation of a preprotein into the matrix
Tim44 appears as a peripheral membrane protein with a large matrix domain and recruits mt-Hsp70 to the site where the preprotein emerges from the Tim17-23 channel. The GTPase is probably bound to the *cis* side of the IM. See the text for detailed explanation.

Two models, the 'Brownian ratchet' [23] and the 'molecular motor' [24], have been advanced to describe how mt-Hsp70 participates in import. Both agree that mt-Hsp70 and ATP hydrolysis in the matrix participate in vectorial movement of preproteins into this compartment. The specifics of how this movement is achieved, however, remain unresolved. The Brownian ratchet mechanism is based on the observation that preproteins *en route* to the matrix can oscillate back and forth within the translocation channels. As a result of Brownian motion, segments of the polypeptide chain in transit periodically become exposed on the matrix side. Subsequent binding of the mt-Hsp70–Tim44 complex to these exposed segments has been proposed to prevent backward movement of the incoming polypeptide chain. A sequence of such binding events converts the random oscillation into unidirectional movement of the polypeptide into the matrix. According to the molecular motor model, binding of the incoming polypeptide chain to the mt-Hsp70–Tim44 complex stimulates ATP hydrolysis by mt-Hsp70, leading to a conformational change in mt-Hsp70 which in turn is used to actively pull a segment of the bound polypeptide chain into the matrix. It is difficult to decide between the two models described above; they may not be mutually exclusive [1,2].

Proteolytic maturation of preproteins in the matrix

For most precursors that are targeted to the matrix, the signal sequence is removed in one step by MPP. The enzyme consists of two subunits (α and β) and both are required for activity. In many cases, the MPP cleavage requires a basic residue (mostly Arg) at position -2 and often also in position -3 with respect to the cleavage site. In yeast, disruption of either of the two subunits of MPP is lethal. Although the targeting function of the signal sequence is not dependent on its removal by MPP, removal of these sequences from precursor proteins might be required either for proper assembly of imported proteins or for their dissociation from other components of the import machinery.

A subset of preproteins undergoes a second processing step: the precursor is first cleaved by MPP to generate an intermediate. The intermediate is subsequently processed to the mature form by the mitochondrial intermediate peptidase, which is also located in the matrix. The dual cleavage site is usually characterized by the motif RX$\uparrow^1$(F/L/I)XX(T/S/G)XXXX$\uparrow^2$, where cleavage ($\uparrow$) 1 represents the MPP site and cleavage 2 represents the mitochondrial intermediate peptidase site [1].

Protein folding in the matrix

Preproteins are translocated across the mitochondrial membranes in an extended conformation. Following import and maturation, these proteins must fold properly into their active conformations and/or assemble into functionally active complexes. As mentioned above, the mt-Hsp70–Tim44 complex binds to preproteins entering into the matrix. The nucleotide-dependent interactions of mt-Hsp70, with Tim44 and also with substrate proteins, is regulated by Mge1, a homologue of bacterial GrpE. Mge1 is an essential protein and serves as a nucleotide-exchange factor. In the simplest case, proteins are released from mt-Hsp70, after which they fold spontaneously. The folding pathway for most proteins, however, is more complex and requires the participation of other chaperones. For example, a co-chaperone Mdj1, the homologue of bacterial DnaJ, co-operates with mt-Hsp70 to prevent misfolding of some imported proteins. Furthermore, a majority of imported proteins fold with the help of Hsp60 and Hsp10, which are the mitochondrial homologues of bacterial GroEL and GroES, respectively. For some proteins, the *cis/trans* isomerization of peptide bonds preceding a prolyl residue is critical for folding, and this is catalysed by peptidyl-prolyl *cis/trans* isomerases [1].

Perspectives

Over the past 20 years we have learned a great deal about mitochondrial biogenesis. However, the field is still wide open with numerous fundamental questions remaining. One area is the identification of new components of the mitochondrial import machinery. Although many Tom and Tim proteins have been reported, the complexity of the mitochondrial two-membrane import machinery makes it likely that other components remain to be identified. For example, GTP plays an essential role in matrix-protein import, its effect is probably mediated by GTP-binding protein(s), and yet to date none have been identified as part of the Tom or Tim machinery. Perhaps the most puzzling area is how the protein-conducting channels work. In mitochondria, the IM must remain impermeable to small molecules. Translocation of proteins across the IM channel therefore must not lead to any concomitant free exchange of ions and small metabolites between the IMS and the matrix, which would be detrimental to mitochondrial functions. How this is achieved is a mystery. Finally, a major challenge lies in using our knowledge of mitochondrial biogenesis for determining the aetiology of, and developing therapies for, some of the diseases related to mitochondrial dysfunction. Several neurodegenerative disorders, ischaemic heart disease, late-onset diabetes and aging all have features related to mitochondrial dysfunction [25].

Summary

- *Mitochondria import most of their proteins from the cytosol. Precursor forms of most matrix proteins as well as some IM and IMS proteins are synthesized on cytoplasmic ribosomes with N-terminal cleavable signal sequences. Many other mitochondrial proteins including IM carrier proteins contain internal targeting sequences.*

- *Three multisubunit translocases, one in the OM and two in the IM, participate in the import process. These translocases co-operate with cytosolic chaperones, chaperone-like soluble proteins in the IMS as well as chaperones in the matrix.*
- *Insertion of carrier proteins into the IM only requires a membrane potential. On the other hand, translocation of preproteins across the IM into the matrix requires (i) a membrane potential, (ii) GTP hydrolysis, which occurs at the outer side of the IM, and (iii) ATP-dependent interactions occurring at the matrix side.*
- *Following import, the cleavable signal sequence of most preproteins is removed in one step by the MPP. In some cases, removal of the signal sequence is achieved in two steps; first by MPP and second by either mitochondrial intermediate peptidase or by IM peptidases.*
- *Imported proteins must be folded properly to perform their functions.*

We thank Norbert Schülke for help with the artwork. D.P. and A.D. are supported by grants GM57067 and DK53953, respectively, from the National Institutes of Health (NIH). D.M.G. is supported by the NIH training grant HL07027.

References

1. Neupert, W. (1997) Protein import into mitochondria. *Annu. Rev. Biochem.* **66**, 863–917
2. Voos, W., Martin, H., Krimmer, T. & Pfanner, N. (1999) Mechanisms of protein translocation into mitochondria. *Biochim. Biophys. Acta* **1422**, 235–254
3. Tokatlidis, K. & Schatz, G. (1999) Biogenesis of mitochondrial inner membrane proteins. *J. Biol. Chem.* **274**, 35285–35288
4. Söllner, T., Griffiths, G., Pfaller, R., Pfanner, N. & Neupert, W. (1989) MOM19, an import receptor for mitochondrial precursor proteins. *Cell* **59**, 1061–1070
5. Kiebler, M., Keil, P., Schneider, H., van der Klei, I.J., Pfanner, N. & Neupert, W. (1993) The mitochondrial receptor complex: a central role of MOM22 in mediating preprotein transfer from receptors to the general insertion pore. *Cell* **74**, 483–492
6. Gratzer, S., Lithgow, T., Bauer, R.E., Lamping, E., Paltauf, F., Kohlwein, S.D., Haucke, V., Junne, T., Schatz, G. & Horst, M. (1995) Mas37p, a novel receptor subunit for protein import into mitochondria. *J. Cell Biol.* **129**, 25–34
7. Söllner, T., Pfaller, R., Griffiths, G., Pfanner, N. & Neupert, W. (1990) A mitochondrial import receptor for the ATP/ADP carrier. *Cell* **62**, 107–115
8. Dietmeier, D., Hönlinger, A., Bömer, U., Dekker, P.J.T., Eckerskorn, C., Lottspeich, F., Kübrich, M. & Pfanner, N. (1997) Tom5 functionally links mitochondrial preprotein receptors to the general import pore. *Nature (London)* **388**, 195–200
9. Kassenbrock, C.K., Cao, W. & Douglas, M.G. (1993) Genetic and biochemical characterization of ISP6, a small mitochondrial outer membrane protein associated with the protein translocation complex. *EMBO J.* **12**, 3023–3034
10. Hönlinger, A., Bömer, U., Alconada, A., Eckerskorn, C., Lottspeich, F., Dietmeier, K. & Pfanner, N. (1996) Tom7 modulates the dynamics of the mitochondrial outer membrane translocase and plays a pathway-related role in protein import. *EMBO J.* **15**, 2125–2137
11. Vestweber, D., Brunner, J., Baker, A. & Schatz, G. (1989) A 42K outer membrane protein is a component of the yeast mitochondrial import site. *Nature (London)* **341**, 205–209

12. Kiebler, M., Pfaller, R., Söllner, T., Griffiths, G., Horstmann, H., Pfanner, N. & Neupert, W. (1990) Identification of a mitochondrial receptor complex required for recognition and membrane insertion of precursor proteins. *Nature (London)* **348**, 610–616
13. Hill, K., Model, K., Ryan, M.T., Dietmeier, K., Martin, F., Wagner, R. & Pfanner, N. (1998) Tom40 forms the hydrophilic channel of the mitochondrial import pore for preproteins. *Nature (London)* **395**, 516–521
14. Künkele, K.-P., Heins, S., Dembowski, M., Nargang, F.E., Benz, R., Thieffry, M., Walz, J., Lill, R., Nussberger, S. & Neupert, W. (1998) The preprotein translocation channel of the outer membrane of mitochondria. *Cell* **93**, 1009–1019
15. Schatz, G. (1997) Just follow the acid chain. *Nature (London)* **388**, 121–122
16. Maarse, A.C., Blom, J., Keil, P., Pfanner, N. & Meijer, M. (1994) Identification of the essential yeast protein MIM17, an integral mitochondrial inner membrane protein involved in protein import. *FEBS Lett.* **349**, 215–221
17. Ryan, K.R., Menold, M.M., Garrett, S. & Jensen, R.E. (1994) SMS1, a high-copy suppressor of the yeast mas6 mutant, encodes an essential inner membrane protein required for mitochondrial protein import. *Mol. Biol. Cell* **5**, 529–538
18. Dekker, P.J., Keil, P., Rassow, J., Maarse, A.C., Pfanner, N. & Meijer, M. (1993) Identification of MIM23, a putative component of the protein import machinery of the mitochondrial inner membrane. *FEBS Lett.* **330**, 66–70
19. Emtage, J.L.T. & Jensen, R.E. (1993) MAS6 encodes an essential inner membrane component of the yeast mitochondrial protein import pathway. *J. Cell Biol.* **122**, 1003–1012
20. Maarse, A.C., Blom, J., Grivell, L.A. & Meijer, M. (1992) MPI1, an essential gene encoding a mitochondrial membrane protein, is possibly involved in protein import into yeast mitochondria. *EMBO J.* **11**, 3619–3628
21. Scherer, P.E., Manning-Krieg, U.C., Jenö, P., Schatz, G. & Horst, M. (1992) Identification of a 45-kDa protein at the protein import site of the yeast mitochondrial inner membrane. *Proc. Natl. Acad. Sci. U.S.A.* **89**, 11930–11934
22. Sepuri, N.B.V., Gordon, D.M. & Pain, D. (1998) A GTP-dependent "push" is generally required for efficient protein translocation across the mitochondrial inner membrane into the matrix. *J. Biol. Chem.* **273**, 20941–20950
23. Schneider, H.-C., Berthold, J., Bauer, M.F., Dietmeier, K., Guiard, B., Brunner, M. & Neupert, W. (1994) Mitochondrial Hsp70/MIM44 complex facilitates protein import. *Nature (London)* **371**, 768–774
24. Glick, B.S. (1995) Can Hsp70 proteins act as force-generating motors? *Cell* **80**, 11–14
25. Wallace, D.C. (1992) Mitochondrial genetics: a paradigm for aging and degenerative diseases. *Science* **256**, 628–632

7

Pore relations: nuclear pore complexes and nucleocytoplasmic exchange

Michael P. Rout*[1] and John D. Aitchison†[2]

*Laboratory of Cellular and Structural Biology, The Rockefeller University, 1230 York Ave, New York, NY 10021, U.S.A., and †Department of Cell Biology, University of Alberta, Edmonton, Alberta, T6G 2H7, Canada

Introduction

One of the main characteristics distinguishing eukaryotes from prokaryotes is that eukaryotes compartmentalize many life processes within membrane-bound organelles. The most obvious of these is the nucleus, bounded by a double-membraned nuclear envelope (NE). The NE thus acts as a barrier separating the nucleoplasm from the cytoplasm. An efficient, regulated and continuous exchange system between the nucleoplasm and cytoplasm is therefore necessary to maintain the structures of the nucleus and the communication between the genetic material and the rest of the cell. The sole mediators of this exchange are the nuclear pore complexes (NPCs), large proteinaceous assemblies embedded within reflexed pores of the NE membranes [1]. While small molecules (such as nucleotides, water and ions) can freely diffuse across the NPCs, macromolecules such as proteins and ribonucleoprotein (RNP) particles are actively transported in a highly regulated and selective manner. Transport through the NPC requires specific soluble factors that recognize transport substrates in either the nucleoplasm or

[1]*e-mail: rout@rockvax.rockefeller.edu*
[2]*e-mail: john.aitchison@ualberta.ca*

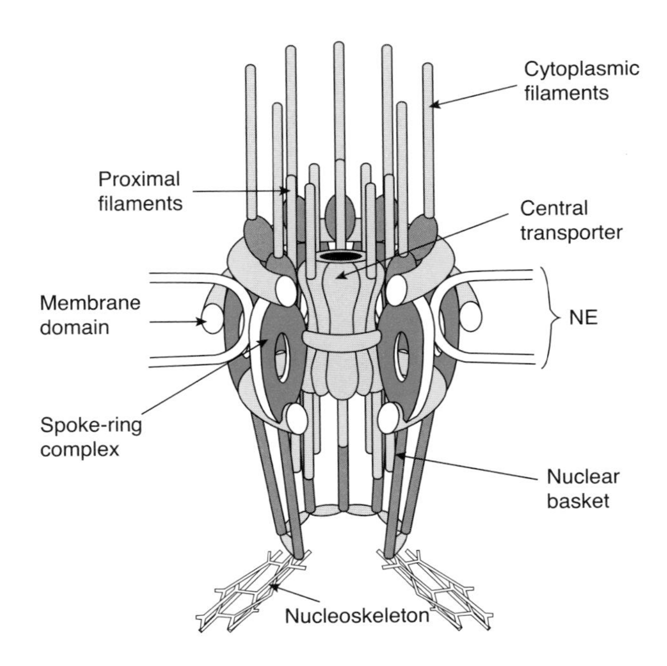

Figure 1. Structure of the NPC
Each NPC is a large proteinaceous assembly composed of a symmetrical cylindrical core, made of eight spokes surrounding a hollow central transporter. Each spoke is composed of several struts and is attached to its neighbours by coaxial rings to form the spoke-ring complex. The NPC is embedded in the NE, and a considerable portion of each spoke traverses the pore membrane and resides in the NE lumen. Peripheral elements include eight cytoplasmic particles and filaments and nuclear filaments which form a basket-like structure attached distally to elements of the nucleoskeleton.

cytoplasm and mediate their transport by docking them to specific components of the NPC [2]. In order to understand how transport works, we must first catalogue the soluble transport factors and NPC components, and then study the details of how they interact.

Despite interesting differences in detail, NPCs from all eukaryotes studied appear to share a common architecture [1,3] (Figure 1). The NPC is comprised of a cylindrical core and a filamentous periphery, and has octagonal rotational symmetry around its cylindrical axis. The core contains a tubular structure (termed the central transporter) surrounded radially by eight spokes interconnected by rings. The core has mirror symmetry in the plane of the NE, and contains both the anchor for the nuclear membrane and the gate for entry to and exit from the nucleus (in the form of the central transporter). Peripheral filaments bristle from the core, projecting into the nucleoplasm and cytoplasm. They appear asymmetric in the plane of the NE; whereas the cytoplasmic filaments spread like a coronet from the cytoplasmic side of the NPC, the nuclear filaments connect at their far ends to form a structure resembling a fish trap or basket.

Nucleoporins, the building blocks of the NPC

The conserved NPC architecture is reflected by the homology between NPC components from different organisms. A variety of techniques have identified numerous NPC component proteins (termed nucleoporins, or NUPs) in several eukaryotes, from yeast to mammals. The amenability of yeast to biochemical and genetic analyses has allowed researchers to use this model organism to identify what may be the entire complement of protein components for the NPC [4–6]. Of course, given the dynamic nature of NPCs and their interactions with soluble factors, it is not always clear what constitutes a NUP. As we shall see, there are NPC components that may associate reversibly with the NPC, and some that are actually found mainly associated with other organelles, and are thus not considered to be NUPs. Nevertheless, some order can be maintained by invoking an operational definition, such that a NUP is considered to be a protein that is mainly associated in a stable fashion with the octagonally symmetric body of the NPC.

Whereas original estimates suggested that the NPC may be composed of as many as 100–200 NUPs [7,8], it now seems that no more than 30 different proteins are needed to build a yeast NPC [4–6]. However, the mass of the NPC is in the range of 50 MDa (yeast) to over 100 MDa (vertebrates) [3,9]. So, for example, if the yeast NPC has only 30 or so proteins, compared with a ribosome of 4 MDa, made of approx. 80 proteins, how is this mass achieved? It seems the answer is in the symmetry. The octagonal and mirror symmetries of the NPC allows NUPs to be present in 8, 16 or even 32 copies, unlike ribosomal proteins, which almost without exception are present in only one copy per ribosome. Furthermore, NUPs can be as large as 350 kDa, and in yeast, for example, have an average molecular mass of approx. 100 kDa, compared with the approx. 25 kDa average molecular mass of a ribosomal protein. Hence, in the NPC, 30 different NUPs with an average mass of approx. 100 kDa, each present in 16 copies, would total 50 MDa (whereas 80 different approx. 25 kDa ribosomal proteins present as one copy each makes the expected approx. 2 MDa, with the remaining 2 MDa supplied by the rRNA).

By localizing NUPs to particular structures and combining this information with data from physical interactions between NUPs, a structural map of the NPC is beginning to emerge [1,5]. NUPs can be divided into three overlapping classes: membrane proteins, core components and components of the peripheral filaments. Pore membrane proteins span the lipid bilayer and, because of their strategic location within the pore, they are presumed to be involved in anchoring the NPC within the NE. Unlike other NUPs, to date there has been no obvious homology found between membrane proteins of yeast and metazoans. This may be due partly to the fact that yeast do not break down the NE and NPCs during mitosis, whereas metaozoans have to rebuild an NPC after each cell division (see below). The NPC core is made of several proteins whose size and abundance allow them to contribute a significant amount of the NPC's mass. Interestingly, only one-third of the yeast NUPs are essential, probably reflecting the multiple contacts each NUP makes with its numerous neighbours [5]. This creates a complex structural framework that is not disrupted by the loss of a single component. As might be expected by the symmetry of the core, most proteins localized within this region seem to be symmetrically disposed on both the cytoplasmic and nucleoplasmic sides [6]. The central transporter lies within the core structure and is also symmetric, as are its components (such as the vertebrate p62/p58/p54 complex found at both ends of the central transporter [10,11], and the similarly located homologous yeast NSP1/NUP57/NUP49 complex [4,12]). Moving away from this central region of the NPC, the structure becomes asymmetric, and some NUPs found distal to the NPC core are found on only one or the other side of the NPC (such as the yeast cytoplasmic NUP159 [13] or the vertebrate nuclear NUP153 [14]).

Building an NPC

Before an NPC can function, it must first be made, and presumably some of the NUPs have a role in NPC assembly. At least in yeast, NPCs appear to be continuously assembled throughout the cell cycle [15]. As leaving a gaping hole in the NE would generally be considered to be a bad thing for a cell, the NPC must punch a hole through both membranes of the NE and insert itself in such a way that neither the nucleoplasm nor the lumen of the endoplasmic reticulum (which is continuous with the NE) leak during this process. Rather than slowly constructing elaborate sealed precursors, it appears that NPCs form by an extremely rapid process without any obvious intermediates. This 'sleight-of-hand' approach may involve the almost simultaneous formation of a correctly sized NE pore, and the insertion of prefabricated subcomplexes which mature into a functional NPC [1,16]. This must be a highly co-operative process, perhaps triggered by local fusion of the inner and outer membranes. In many metazoans, the NE is reversibly disassembled during mitosis, such that they also undergo a round of NPC assembly at telophase. Although it is not clear that the processes of mitotic NPC reassembly are the same as those at interphase, it is presumed that they share similar mechanisms. As a result, mitotic NPC reassembly has been the system of choice to study NPC biogenesis. NPCs probably assemble from the membrane pore proteins inwards [17]. The morphology of the mitotic NPC reassembly intermediates suggests an ordered hierarchical assembly [16,17]. The inner and outer membranes of the NE first fuse locally to form a small pore, which becomes rapidly filled with rings and spokes. This is then followed by the addition of more peripheral rings, and the peripheral filaments.

At least in some cases, NPC components exist as preassembled subcomplexes in the cytoplasm [1,7,16,18], suggesting that the rapidity of NPC assembly owes a lot to its construction from a relatively small number of prefabricated parts. In fact, we might speculate that the main function of some NUPs is to direct NPC assembly. Gp210, for example, would be such a candidate, perhaps required for mitotic reassembly of NPCs; it is conspicuously absent from yeast, which do not disassemble their nuclei at any point in the cell cycle. In addition, non-NUP proteins may be recruited from other parts of the cell to help during assembly, before becoming incorporated into the NPC. This role has been suggested for Sec13, a protein normally found in cytoplasmic coat protein-containing (COP II) vesicles trafficking between the endoplasmic reticulum and Golgi complex, but which was also found in a complex with numerous NUPs [19,20]. As most Sec13 is associated with the endoplasmic reticulum and its vesicles rather than the NPCs, it is an example of an NPC component that is not a NUP. Its function in forming coats around vesicles may be borrowed to help stabilize the reflexed membrane of the nascent nuclear pore, or recruit other proteins to promote fusion during NPC assembly, thus explaining its dual localization [19].

Association of the NPC with adjacent structures

The NPC cannot be considered as an isolated structure, as it interacts with components of its surroundings, including the NE, the nucleoplasm and the cytoplasm. As the whole NPC can move rapidly in the plane of the NE [21,22], these interactions are labile. Such multiple interactions are reflected in individual NUPs. For example, NUP153 has different domains that bind transport factors from the nucleoplasm and cytoplasm, attach it to the NPC, and connect the NPC to the adjacent filamentous nuclear structures [23,24]. Interactions with the filamentous nuclear lamina, which runs in a layer beneath and parallel with the NE, are probably important for maintaining the normal spacing of metazoan NPCs, as disruptions of the lamina can lead to abnormal NPC distributions [25]. Another set of filamentous proteins associated with NPCs is the Tpr family. Though not NUPs, the Tpr homologues extend from the tip of the nuclear basket to form a network that interconnects adjacent NPCs and extends a considerable distance into the nucleoplasm [26,27]. Indeed, this family represents perhaps the best candidates yet for major components of the nucleoskeleton (the nuclear analogue of the cytoskeleton). Because removal of the yeast Tpr homologues decreases the efficiency of nucleocytoplasmic transport, and interactions between Tpr and transport factors have been found *in vitro*, it has been proposed that these proteins act as 'tracks' to guide the movement of nucleocytoplasmic transport between the NPC and deep within the nuclear interior [24,26,28].

The mechanism of transport

Obviously the main function of the NPC, once assembled, is to mediate nucleocytoplasmic exchange; both the passive diffusion of small molecules, and active bi-directional macromolecular transport. Indeed, the assembled membrane and core structures can, in one sense, be considered a framework providing the correct positioning of the NUPs that mediate transport. Transport cargoes are generally recognized first by transport factors in the nucleoplasm or cytoplasm. The transport-factor–cargo complex then docks to the peripheral filaments before translocating through the central transporter, to be released on the other side of the NPC [2]. The components of nucleocytoplasmic transport can therefore be separated into two classes: a stationary phase, consisting of components of the NPC, and a mobile phase of transport factors. Many of these transporters are part of a structurally related family of proteins, collectively termed the karyopherins or Kaps [2,29]. Individual members of the family have numerous alternative names (for example importin, transportin, exportin), as discussed by Barry and Wente in Chapter 8 in this volume.

In the soluble phase, the directionality of transport is determined by where karyopherins load and release their cargoes, which in turn has been shown to be dependent on their interaction with the small Ras-like GTPase Ran. As Ran is maintained in its GTP-bound form in the nucleus and in its GDP-bound form in the cytoplasm, karyopherins can sense their location through Ran, and bind or release their cargoes appropriately (see Figure 2 and legend for details). As the hydrolysis of GTP by Ran releases energy, it seems that Ran not only confers the direction of transport but also powers it [2].

Although we have referred to the NPC as the stationary phase of transport, it is clear that NPCs are in fact very dynamic, and large morphological changes have been observed during transport. One of the most striking examples of this must be the opening of the nuclear basket to accommodate large ribonucleoprotein particles as they unwind through the central transporter [30]. Another example is the central transporter itself, whose structure has been studied at high resolution and found to exist in a number of different conformations. It seems that the central transporter represents the 'transport gate'. In its 'resting' conformation, it has a central hole of approx. 9 nm which prevents the passive diffusion of large molecules across the NPC while permitting the free diffusion of small molecules. However, during the active transport of macromolecules, the central transporter appears to dilate to allow the passage of the transporting materials through it [31]. Indeed, the tube of the central transporter presents a significant hindrance to the free exchange of macromolecules across the NPC. Thus diffusion through the constricted central transporter is an entropically unfavourable process for macromolecules, which therefore tend to be excluded from this region [32].

However, analysis of the NPC components has provided a big clue as to how this barrier may be overcome, and thus how transport through the NPC may occur. Almost half of the known NPC components contain binding sites for numerous transport factors. Large numbers of these binding sites are positioned strategically throughout the NPC: from its cytoplasmic tip, through the

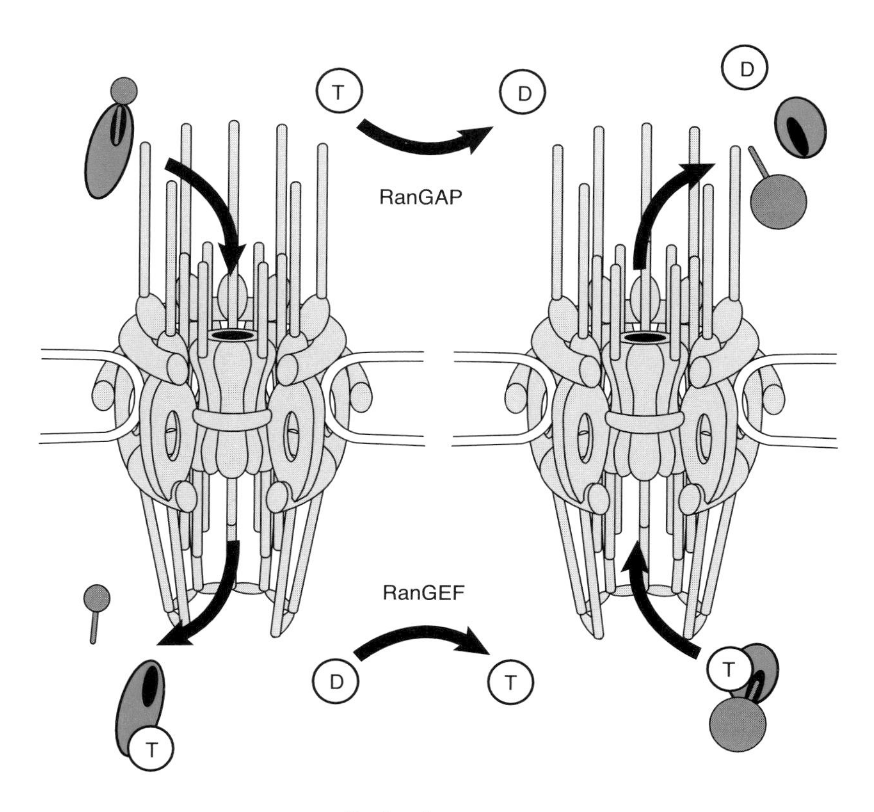

Figure 2. Directional transport is controlled by the interaction of karyopherins with Ran-GTP, NUPs and substrates
Importers release their substrates when they interact with Ran-GTP, while exporters bind Ran-GTP in order to bind their substrates. On the other hand, if Ran hydrolyses its GTP (T) to GDP (D), importers can bind their cargo, whereas exporters release theirs [2,46]. In the nucleus, Ran is maintained in its GTP-bound state by the nuclear-restricted GTP-exchange factor RCC1 [46], whereas the localization of the Ran-GTPase-activating protein (RanGAP) to the cytoplasm and the cytoplasmic filaments of the NPC [47,48] ensures that cytoplasmic Ran hydrolyses its GTP and so is maintained in its GDP-bound form. By so compartmentalizing the modulators of Ran, the cell maintains a gradient of Ran-GTP across the NE. Thus substrates to be imported bind to their transport factors in the cytoplasm in the presence of Ran-GDP, but when this complex meets Ran-GTP in the nucleoplasm the switch is pulled; Ran-GTP binds to the transporter, changing its conformation and causing cargo release. In contrast, as an export complex reaches the cytoplasm, RanGAP stimulates GTP hydrolysis and the cargo is released. Ran is the only known energy-utilizing transport factor, and so the energy driving all nucleocytoplasmic transport may come from this gradient of Ran-GTP [46]. RanGEF, Ran guanine nucleotide-exchange factor.

central transporter to its nuclear tail, and beyond. Most, if not all, of these are distributed along filamentous structures. The best-studied karyopherin-binding sites are the peptide repeat motifs (of Phe-Gly) present in nearly half of the known NUPs [4,5]. In one NUP these repeats have been shown directly to form filaments [33], and numerous other repeat-motif-containing NUPs have been localized to filamentous NPC structures [12,16]. Although there is only minimum amino acid sequence conservation in the repeat motifs between presumed homologues from different species [4,5,8], there may be a high degree of functional conservation as the position of these homologues within the NPC is well conserved (e.g. NSP1/NUP57/NUP49 compared with p62/p58/p54, above). The Brownian motion of these putative filamentous proteins could help to exclude the non-specific diffusion of large molecules across the central tube by a process termed 'entropic exclusion' [34]. In this way, they may contribute to the NPC's apparent resting diameter of approx. 9 nm. On the other hand, these NUPs contain an abundance of binding sites, surrounding the central tube, which would provide a means to recruit transport-factor–cargo complexes to the mouth of the central tube. This would, by contrast, promote the specific diffusion of transport-factor–cargo complexes through the central tube, between the binding sites found on both the nuclear and cytoplasmic sides of the NPC (Figure 3). The NPC would thus effectively be a 'virtual gate'; as proteins that bind the NPC would pass the diffusion barrier of the central channel much more freely than those that do not, gating selectivity would be achieved without necessarily invoking a gate composed of any moving parts. Thus the observed dilation in the NPC may be a consequence of cargo movement.

Taken together, it would seem that NPC-mediated nucleocytoplasmic transport is based on essentially three steps (Figure 3). The first is that the reversible binding of transport factors to the large number of NPC-binding sites would encourage the free diffusion of transport-factor–cargo complexes between both faces of the NPC. This is then followed by a second step, in which the transport-factor–cargo complexes would move preferentially to asymmetric binding sites on the same side of the NPC as their ultimate destination. The particular Ran-bound state of the karyopherin would determine the direction of this step. Thus importers, not bound to Ran, would have their highest affinity for the docking sites on the peripheral nuclear basket. This is consistent with the observed accumulation of a Ran-binding-deficient karyopherin mutant on the nuclear side of the NPC [35]. Similarly Ran-bound export factors would preferentially jump to binding sites on the peripheral cytoplasmic filaments; for example, whereas the export factor Crm1 binds to several repeat-containing NUPs, when complexed to Ran-GTP it binds preferentially to the cytoplasmic NUP214 [36]. The third step involves either Ran-GTP binding or GTP hydrolysis (depending on the direction of transport), which leads to displacement of the cargo from the carrier and the carrier from the NPC [2,29]. This, being essentially irreversible, terminates the transport reaction and ensures the overall directionality of transport. It can readily be seen from Figure 3 how import factors may now be recycled out of the nucle-

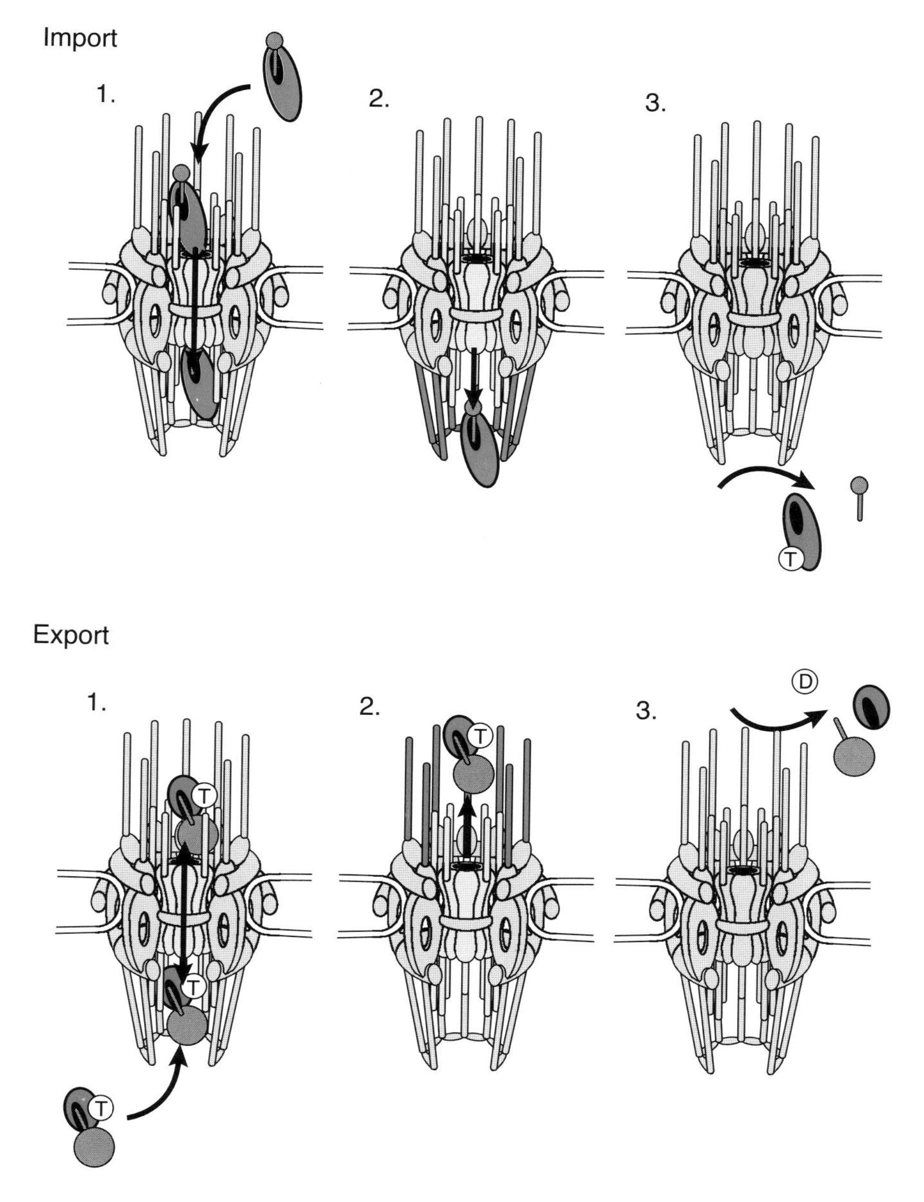

Figure 3. A model for nuclear import and export through the NPC
Karyopherins pick up their cargo in either the nucleus or cytoplasm, and bind to the NPC by interacting reversibly with (repeat-containing) NUPs (step 1). Directional movement is ensured by movement to higher-affinity terminal docking sites on the NPC side opposite that from which the karyopherins started (step 2), followed by Ran-mediated release at these sites (step 3). See text for details. D, RanGDP; T, RanGTP.

us, in a manner analogous to the transport of export factors without cargo. Similarly, recycling of export factors may be analogous to cargo-less import.

The energy driving nucleocytoplasmic transport may therefore be provided by the hydrolysis of Ran-bound GTP during each transport and recycling round.

Although the NPC contains an abundance of repeats, and different karyopherins can recognize the same NUP, different karyopherins have strong preferences for particular repeat motif classes. In one case it has been shown that a particular karyopherin (Kap121) appears to utilize a particular repeat-containing docking site within NUP53. Kap121 is the only karyopherin detected in association with NUP53, and accumulation of Kap121 at the NPC is specifically reduced in the absence of NUP53 [36a]. It therefore seems that one reason for the large number of repeat-motif NUPs is to mediate separate pathways across the NPC for different kinds of karyopherins, reducing competition between these pathways at the NPC and creating a potential for differential control [8,29,36a]. As deletion of Kap121 is lethal, but the deletion of its favoured docking site is not [36a], another important principle seems to be that these pathways are not totally separate; they can overlap if necessary, with each karyopherin therefore having a choice of more or less favoured pathways across the NPC.

The many different classes of repeat motifs may also reflect the fact that there may be nuclear transport factors other than karyopherins, and they also use them as NPC docking sites. Thus the TAP/p15 complex (Mex67/Mtr2 in yeast), which is implicated in RNA export, binds specifically to NUP repeat motifs during its passage across the NPC [37,38]. Similarly Ntf2, which is suggested to mediate Ran exchange between the nucleoplasm and cytoplasm but also seems involved in a number of other processes, binds reversibly to repeat-motif domains [2,39,40]. Finally, different classes of repeat-motif NUPs may be bound at different stages of transport. In particular, while the symmetrically localized NUPs are docked in the initial stages of transport, the terminal reaction of each transport cycle occurs on one of the distal asymmetrical repeat-motif NUPs [2,36]. It may be that the particular order of the binding sites within the NPC helps to correctly guide or channel the transport factors as they cross the NPC.

In addition to repeat motifs, additional binding sites exist within NUPs. For example, binding sites for the energy-providing protein Ran abound in vertebrate NUPs, both in the form of the conserved Ran-binding domain and in the more recently characterized Ran-binding zinc-finger motifs [2,41–43]. This may serve to increase the concentration of Ran at the NPC and hence improve the efficiency of the transport-termination steps. They may also be involved in mediating an exchange of Ran across the NPC, and in maintaining the correct balance of Ran-GDP in the cytoplasm and Ran-GTP in the nucleoplasm. This latter possibility is supported by the presence of binding sites for RanGAP1 (Ran-GTPase-activating protein 1, the protein that maintains the GDP form of Ran) on the cytoplasmic filaments of vertebrate NPCs [2] (see also Figure 2). Similarly, mRNP41 (Rae1 or Gle2 in yeast), which is also implicated in RNA export, cycles on and off a unique binding site [44]. The Gle2-

binding domain is essential, but autonomous, as it can be moved from one NUP to another [45]. The list of binding sites may not end there; NUP358, the current record holder for the number and variety of such binding sites, contains a coiled-coil domain (anchoring it to the NPC), four Ran-binding domains, a zinc-finger region (also for binding Ran), numerous FG repeats (for binding karyopherins, Ntf2 and the like), two RanGAP1-binding repeats and a cyclophilin A (prolyl isomerase) homology domain [41].

Perspectives

Over the past few years, there have been tremendous advances in our understanding of the molecular mechanisms of nuclear transport. Most of our new understanding has focused on the soluble transport factors, so the challenge now is to determine how these factors interface with the NPC to mediate directional transport. Although we have presented a model for how transport may occur through the NPC, based on our current knowledge it is important to realize that this simple scheme is speculative and leaves many questions unanswered. In particular, does Ran have a role within the NPC, in addition to the terminal reactions? What is the role of Ran-binding sites within specific NUPs? (Not to mention the plethora of other binding sites on proteins like NUP358). How does the NPC change its conformation during transport? How many contacts do cargo–carrier complexes make during transport and how do these translate into directional movement through the NPC? An important step towards answering these questions will be to complete a map of NUPs in the NPC, and to determine the proteins they contact both temporally and spatially during transport and NPC biogenesis.

Another rapidly expanding area of research is that of nucleocytoplasmic transport regulation. Although in most cases this is achieved by controlling the affinity of a transport substrate for its transport factor [2], intriguing evidence is emerging that NUPs are also regulated. In yeast, NUP53 is phosphorylated specifically during mitosis [36a]. This correlates with a decrease in the binding of the NPC to Kap121, which prefers NUP53 as a docking site. Thus it appears that phosphorylation is used to control the affinity of a NUP's docking site for its transport factors. It has also been shown that many of the repeat-motif NUPs are modified by O-linked glycosylations [1,16], although the reasons for this are still unclear.

Studies on the function of the NPC, and the roles of individual NUPs in regulated nuclear transport, NPC assembly, gene expression and the maintenance of cellular structure are all expected to be very fruitful areas of future research, neatly complementing the parallel work on the interaction between transport factors and substrates.

Summary

- *NPCs are the sole sites of exchange between the nucleus and cytoplasm.*
- *A large family of transport factors carry cargo between the nucleus and cytoplasm through the NPC.*
- *The NPC is a huge symmetric octagonal structure comprised of dozens of NUPs.*
- *NUPs make many contacts with surrounding structures, including the NE, the cytoplasm and nuclear interior.*
- *A subset of NUPs contain repeated peptide motifs that serve as docking sites for transport factors.*
- *The directionality of transport is determined by the transport factor, and its interactions with the small GTPase Ran and NUPs.*
- *Very little is known about how the NPC mediates transport, NPC assembly and the NPC's role in regulating transport, but these areas of research are beginning to emerge.*

We are grateful to Colin Dingwall and Rick Wozniak for critical reading of the manuscript, and to Beth Hatton for excellent secretarial assistance. M.P.R. is supported by the Rita Allen Foundation and the Irma Hirschl Trust and J.D.A. by the Medical Research Council of Canada and Alberta Heritage Foundation for Medical Research. Apologies to those whose original work could not be cited due to space limitations; readers are encouraged to refer to the original publications.

References

1. Davis, L.I. (1995) The nuclear pore complex. *Annu. Rev. Biochem.* **64**, 865–896
2. Mattaj, I.W. & Englmeier, L. (1998) Nucleocytoplasmic transport: the soluble phase. *Annu. Rev. Biochem.* **67**, 265–306
3. Yang, Q., Rout, M.P. & Akey, C.W. (1998) Three-dimensional architecture of the isolated yeast nuclear pore complex: functional and evolutionary implications. *Mol Cell.* **1**, 223–234
4. Doye, V. & Hurt, E. (1997) From nucleoporins to nuclear pore complexes. *Curr. Opin. Cell Biol.* **9**, 401–411
5. Fabre, E. & Hurt, E. (1997) Yeast genetics to dissect the nuclear pore complex and nucleocytoplasmic trafficking. *Annu. Rev. Genet.* **31**, 277–313
6. Rout, M.P., Aitchison, J.D., Suprapto, A., Hjertaas, K., Zhao, Y. & Chait, B.T. (2000) The yeast nuclear pore complex: composition, architecture and mechanism. *J. Cell Biol.* **148**, 635–651
7. Forbes, D.J. (1992) Structure and function of the nuclear pore complex. *Annu. Rev. Cell Biol.* **8**, 495–527
8. Rout, M.P. & Wente, S.R. (1994) Pores for thought: nuclear pore proteins. *Trends Cell Biol.* **4**, 357–365
9. Reichelt, R., Holzenburg, A., Buhle, Jr, E.L., Jarnik, M., Engel, A. & Aebi, U. (1990) Correlation between structure and mass distribution of the nuclear pore complex and of distinct pore complex components. *J. Cell Biol.* **110**, 883–894
10. Grote, M., Kubitscheck, U., Reichelt, R. & Peters, R. (1995) Mapping of nucleoporins to the center of the nuclear pore complex by post-embedding immunogold electron microscopy. *J. Cell Sci.* **108**, 2963–2972

11. Guan, T., Muller, S., Klier, G., Pante, N., Blevitt, J.M., Haner, M., Paschal, B., Aebi, U. & Gerace, L. (1995) Structural analysis of the p62 complex, an assembly of O-linked glycoproteins that localizes near the central gated channel of the nuclear pore complex. *Mol. Biol. Cell* **6**, 1591–1603
12. Fahrenkrog, B., Hurt, E.C., Aebi, U. & Pante, N. (1998) Molecular architecture of the yeast nuclear pore complex: localization of Nsp1p subcomplexes. *J. Cell Biol.* **143**, 577–588
13. Kraemer, D.M., Strambio-de-Castillia, C., Blobel, G. & Rout, M.P. (1995) The essential yeast nucleoporin NUP159 is located on the cytoplasmic side of the nuclear pore complex and serves in karyopherin-mediated binding of transport substrate. *J. Biol. Chem.* **270**, 19017–19021
14. Sukegawa, J. & Blobel, G. (1993) A nuclear pore complex protein that contains zinc finger motifs, binds DNA, and faces the nucleoplasm. *Cell* **72**, 29–38
15. Winey, M., Yarar, D., Giddings, Jr, T.H. & Mastronarde, D.N. (1997) Nuclear pore complex number and distribution throughout the *Saccharomyces cerevisiae* cell cycle by three-dimensional reconstruction from electron micrographs of nuclear envelopes. *Mol. Biol. Cell* **8**, 2119–2132
16. Bastos, R., Pante, N. & Burke, B. (1995) Nuclear pore complex proteins. *Int. Rev. Cytol.* **162B**, 257–302
17. Gant, T.M., Goldberg, M.W. & Allen, T.D. (1998) Nuclear envelope and nuclear pore assembly: analysis of assembly intermediates by electron microscopy. *Curr. Opin. Cell Biol.* **10**, 409–415
18. Macaulay, C., Meier, E. & Forbes, D.J. (1995) Differential mitotic phosphorylation of proteins of the nuclear pore complex. *J. Biol. Chem.* **270**, 254–262
19. Siniossoglou, S., Wimmer, C., Rieger, M., Doye, V., Tekotte, H., Weise, C., Emig, S., Segref, A. & Hurt, E.C. (1996) A novel complex of nucleoporins, which includes Sec13p and a Sec13p homolog, is essential for normal nuclear pores. *Cell* **84**, 265–275
20. Fontoura, B.M., Blobel, G. & Matunis, M.J. (1999) A conserved biogenesis pathway for nucleoporins: proteolytic processing of a 186-kilodalton precursor generates nup98 and the novel nucleoporin, nup96. *J. Cell Biol.* **144**, 1097–1112
21. Belgareh, N. & Doye, V. (1997) Dynamics of nuclear pore distribution in nucleoporin mutant yeast cells. *J. Cell Biol.* **136**, 747–759
22. Bucci, M. & Wente, S.R. (1997) *In vivo* dynamics of nuclear pore complexes in yeast. *J. Cell Biol.* **136**, 1185–1199
23. Enarson, P., Enarson, M., Bastos, R. & Burke, B. (1998) Amino-terminal sequences that direct nucleoporin nup153 to the inner surface of the nuclear envelope. *Chromosoma* **107**, 228–236
24. Shah, S., Tugendreich, S. & Forbes, D. (1998) Major binding sites for the nuclear import receptor are the internal nucleoporin Nup153 and the adjacent nuclear filament protein Tpr. *J. Cell Biol.* **141**, 31–49
25. Lenz-Bohme, B., Wismar, J., Fuchs, S., Reifegerste, R., Buchner, E., Betz, H. & Schmitt, B. (1997) Insertional mutation of the *Drosophila* nuclear lamin Dm0 gene results in defective nuclear envelopes, clustering of nuclear pore complexes, and accumulation of annulate lamellae. *J. Cell Biol.* **137**, 1001–1016
26. Cordes, V.C., Reidenbach, S., Rackwitz, H.R. & Franke, W.W. (1997) Identification of protein p270/Tpr as a constitutive component of the nuclear pore complex-attached intranuclear filaments. *J. Cell Biol.* **136**, 515–529
27. Strambio-de-Castillia, C., Blobel, G. & Rout, M.P. (1999) Proteins connecting the nuclear pore complex with the nuclear interior. *J. Cell Biol.* **144**, 839–855
28. Bangs, P., Burke, B., Powers, C., Craig, R., Purohit, A. & Doxsey, S. (1998) Functional analysis of tpr: identification of nuclear pore complex association and nuclear localization domains and a role in mRNA export. *J. Cell Biol.* **143**, 1801–1812
29. Wozniak, R.W., Rout, M.P. & Aitchison, J.D. (1998) Karyopherins and kissing cousins. *Trends Cell Biol.* **8**, 184–188
30. Kiseleva, E., Goldberg, M.W., Daneholt, B. & Allen, T.D. (1996) RNP export is mediated by structural reorganization of the nuclear pore basket. *J. Mol. Biol.* **260**, 304–311
31. Akey, C.W. (1990) Visualization of transport-related configurations of the nuclear pore transporter. *Biophys. J.* **58**, 341–355

32. Feldherr, C.M. & Akin, D. (1997) The location of the transport gate in the nuclear pore complex. *J. Cell Sci.* **110**, 3065–3070
33. Buss, F., Kent, H., Stewart, M., Bailer, S.M. & Hanover, J.A. (1994) Role of different domains in the self-association of rat nucleoporin p62. *J. Cell Sci.* **107**, 631–638
34. Brown, H.G. & Hoh, J.H. (1997) Entropic exclusion by neurofilament sidearms: a mechanism for maintaining interfilament spacing. *Biochemistry* **36**, 15035–15040
35. Gorlich, D., Pante, N., Kutay, U., Aebi, U. & Bischoff, F.R. (1996) Identification of different roles for RanGDP and RanGTP in nuclear protein import. *EMBO J.* **15**, 5584–5594
36. Kehlenbach, R.H., Dickmanns, A., Kehlenbach, A., Guan, T. & Gerace, L. (1999) A role for RanBP1 in the release of CRM1 from the nuclear pore complex in a terminal step of nuclear export. *J. Cell Biol.* **145**, 645–657
36a. Marelli, M., Aitchison, J.D. & Wozniak, R.W. (1998) Specific binding of the karyopherin kap121p to a subunit of the nuclear pore complex containing nup53p, nup59p, and nup170p. *J. Cell Biol.* **143**, 1813–1830
37. Gruter, P., Tabernero, C., von Kobbe, C., Schmitt, C., Saavedra, C., Bachi, A., Wilm, M., Felber, B.K. & Izaurralde, E. (1998) TAP, the human homolog of Mex67p, mediates CTE-dependent RNA export from the nucleus. *Mol. Cell* **1**, 649–659
38. Katahira, J., Sträßer, K., Podtelejnikov, A., Mann, M., Jung, J.U. & Hurt, E. (1999) The Mex67p-mediated nuclear mRNA export pathway is conserved from yeast to human. *EMBO J.* **18**, 2593–2609
39. Nehrbass, U. & Blobel, G. (1996) Role of the nuclear transport factor p10 in nuclear import. *Science* **272**, 120–122
40. Ribbeck, K., Lipowsky, G., Kent, H.M., Stewart, M. & Gorlich, D. (1998) NTF2 mediates nuclear import of Ran. *EMBO J.* **17**, 6587–6598
41. Wu, J., Matunis, M.J., Kraemer, D., Blobel, G. & Coutavas, E. (1995) Nup358, a cytoplasmically exposed nucleoporin with peptide repeats, Ran-GTP binding sites, zinc fingers, a cyclophilin A homologous domain, and a leucine-rich region. *J. Biol. Chem.* **270**, 14209–14213
42. Nakielny, S., Shaikh, S., Burke, B. & Dreyfuss, G. (1999) Nup153 is an M9-containing mobile nucleoporin with a novel Ran-binding domain. *EMBO J.* **18**, 1982–1995
43. Yaseen, N.R. & Blobel, G. (1999) Two distinct classes of ran-binding sites on the nucleoporin nup-358. *Proc. Natl. Acad. Sci. U.S.A.* **96**, 5516–5521
44. Murphy, R., Watkins, J.L. & Wente, S.R. (1996) GLE2, a *Saccharomyces cerevisiae* homologue of the *Schizosaccharomyces pombe* export factor RAE1, is required for nuclear pore complex structure and function. *Mol. Biol. Cell* **7**, 1921–1937
45. Bailer, S.M., Siniossoglou, S., Podtelejnikov, A., Hellwig, A., Mann, M. & Hurt, E. (1998) Nup116p and nup100p are interchangeable through a conserved motif which constitutes a docking site for the mRNA transport factor gle2p. *EMBO J.* **17**, 1107–1119
46. Moore, M.S. (1998) Ran and nuclear transport. *J. Biol. Chem.* **273**, 22857–22860
47. Matunis, M.J., Coutavas, E. & Blobel, G. (1996) A novel ubiquitin-like modification modulates the partitioning of the Ran-GTPase-activating protein RanGAP1 between the cytosol and the nuclear pore complex. *J. Cell Biol.* **135**, 1457–1470
48. Mahajan, R., Delphin, C., Guan, T., Gerace, L. & Melchior, F. (1997) A small ubiquitin-related polypeptide involved in targeting RanGAP1 to nuclear pore complex protein RanBP2. *Cell* **88**, 97–107

8

Nuclear transport: never-ending cycles of signals and receptors

Dianne M. Barry and Susan R. Wente[1]

Department of Cell Biology and Physiology, Washington University School of Medicine, Box 8228, 660 S. Euclid Ave., St. Louis, MO 63110, U.S.A.

Introduction

In eukaryotic cells, the nuclear envelope spatially and temporally segregates gene expression into two sequential processes. RNA synthesis and ribosome assembly take place within the nucleus, whereas translation is a cytoplasmic event. Selective nucleocytoplasmic exchange of molecules is mediated by the co-ordinated efforts of distinct soluble proteins and components of the nuclear pore complex (NPC). As discussed in greater detail elsewhere (see Chapter 7 in this volume, by Rout and Aitchison), NPCs are comprised of at least 30 distinct proteins termed nucleoporins, or NUPs. Embedded in the nuclear envelope, these large NPC structures provide an aqueous channel through which transport substrates can navigate. Macromolecules targeted for translocation contain specific consensus sequences referred to as nuclear localization sequences (NLSs) or nuclear export sequences (NESs) depending on the directionality of transport. Individual members of a transport receptor superfamily bind to the respective consensus sites, and accumulating evidence suggests a model in which translocating hetero-oligomeric complexes move through the NPC by transiently associating with particular nucleoporins. Once transported to the opposite face of the NPC, the complex disassembles;

[1]*To whom correspondence should be addressed (e-mail: swente@cellbio.wustl.edu).*

the substrates are localized to their sites of action and the transport factors are recycled to undergo continuous rounds of transport.

Signals that direct transport

Nuclear localization is now known to be a signal-driven event. However, notable distinctions exist between the mechanism of nuclear transport and that for protein import across the membranes of the endoplasmic reticulum, the mitochondria and the chloroplasts (see Chapters 1, 5 and 6 in this volume, by Meacock et al., by Schnell, and by Pain et al., respectively). First, nucleocytoplasmic transport is bi-directional in that macromolecules move both into and out of the nucleus. Secondly, proteins transported between the nucleus and cytoplasm are folded in their native and functional conformations; in many other types of organelle transport, unfolded protein substrates interact with transport machinery. Thirdly, in contrast to the cleaved signal sequences that direct proteins to some other subcellular compartments, the signals contained within proteins targeted for nucleocytoplasmic transport are not removed during transit. Fourthly, the NLS and NES signals can be present anywhere within the primary amino acid sequence. In addition, many different NLS and NES signal types have been identified that exhibit a striking and surprising diversity in amino acid composition. These distinct classes of transport signals reflect multiple saturable, non-competing nuclear import and export pathways that are mediated by alternative transport machinery. Finally, recognition of some nuclear-transport signals is regulated during development and in response to environmental stimuli.

Nuclear localization

Following translation in the cytoplasm, proteins are incorporated into specific subcellular compartments. Although proteins smaller than approx. 40 kDa may diffuse passively between the nucleus and cytoplasm, larger macromolecules must be transported actively across the nuclear envelope. In the early 1980s, a short stretch of basic amino acids in the simian virus 40 large T antigen was identified that governs the nuclear localization of this protein and which is sufficient to target an otherwise cytoplasmic protein to the nucleus [1]. This NLS is often referred to as the 'classical' NLS (cNLS; Table 1). Interestingly, another class of NLS was coincidentally characterized that is strongly related to the cNLS. Instead of one binding site (monopartite) for the transport receptor, two sites (bipartite) are utilized [2], presumably to stabilize the interaction between the substrate and the transport machinery [3]. Up to 50% of nuclear proteins contain a bipartite cNLS [2].

Because the import of certain proteins and U small nuclear ribonucleoprotein particles (snRNPs) does not compete with the cNLS pathway [4–6], alternative import pathways were suggested and later shown to exist. A distinct NLS, termed M9, is necessary and sufficient for the nuclear import of het-

Table 1. *Cis*-acting transport signals in proteins

References for the transport receptors are given in Table 2. For arginine-rich NLS (R-NLS), the sequence shown is for HIV-1 Rev, and the underlined sequence is identical to that of the HIV-1 Tat protein. Multiple names for a single transport receptor reflect its coincidental identification by several laboratories. SV40, simian virus 40; hnRNP, heterogenous nuclear RNP.

Type	Prototypical protein(s)	Consensus sequence	Transport receptor(s)	Reference for signal
Import mediators				
cNLS				
Monopartite	SV40 large T antigen	PKKKRKV	Importin/karyopherin $\alpha/\beta 1$	[1]
Bipartite	Nucleoplasmin, hnRNP K	$(K/R)_2X_{10-12}(K/R)_3$	Importin/karyopherin $\alpha/\beta 1$	[2,7]
Ribo NLS	Ribosomal L25	MAPSAKATAAKKAVVKGTNGKKALKV- -RTSATFRLPKTLKLARAP	Kap123, Kap121	[23]
M9 NLS	hnRNP A1	NQSSNFGPMKGGNFGGRSSGPYGGGG- -QYFAKPRNQGGY	Transportin	[6]
KNS NLS	hnRNP K	GFSADETWDSAIDTWSPSEWQMAY	?	[7]
R-NLS	HIV-1 Rev and Tat HTLV-1 Rex	RQARRNRRRRWR*	?	[8,9]
Export mediators				
Leucine-rich NES	PKI, HIV-1 Rev	$LX_{2-3}LX_{2-3}LXL/I$	Crm1/Xpo1/exportin	[6]
M9 NES	hnRNP A1	As for M9 NLS	?	[6]

erogenous nuclear RNP (hnRNP) protein A1, an mRNA-binding protein [6]. The M9 NLS is considerably longer and bears no primary sequence resemblance to the cNLSs (Table 1). Interestingly, another mRNA-binding protein, hnRNP K, has two types of NLS: a bipartite cNLS and a novel NLS called KNS [7] (Table 1). Each NLS in hnRNP K independently promotes nuclear import. This finding suggests that at least some cellular proteins may access multiple different import pathways. It is important to note that NLSs have not been identified for all nuclear proteins, indicating that additional pathways probably exist for nuclear import.

It has long been known that viruses infect cells and subjugate cellular machinery for the purpose of expressing viral gene products, replicating the viral genome and ensuring viral proliferation. By pirating the existing cellular transport pathways, viral proteins accomplish these tasks. Very recently, a novel NLS was reported in the human immunodeficiency virus (HIV)-1 Rev protein as well as the Rex protein of human T-cell leukemia virus type 1 (HTLV-1) [8,9]. In contrast to the cellular NLSs, these viral sequences are rich in arginines (R-NLS; Table 1). It is possible that endogenous cellular proteins also contain this type of NLS.

Nuclear export

More than 10 years after the first report of a NLS, the hunt for NESs revealed a leucine-rich (LR) stretch of eight amino acids in the HIV-1 protein Rev and in the cellular protein, cAMP-activated protein-kinase inhibitor (PKI) [6]. The NES is necessary and sufficient to promote nuclear export. Several groups have since reported the identification of this same LR-NES motif in shuttling (TFIIIA), NPC-associated (yeast Gle1 and Mex67p) and predominantly cytoplasmic [Ran-binding protein 1 (RanBP1)/Yrb1] proteins (Table 1) [4,6,10,11]. Interestingly, the same group that pinned down M9 as an NLS reported that M9 also acts as an NES [6]. Other NLSs, such as the KNS and cNLS, do not confer export activity. This suggests that the same amino acid sequence can endow a single protein with either export or import properties, depending on the environmental context of the protein.

RNA signals

Proteins are not the only molecules that move between the nucleus and the cytoplasm; nuclear export of all classes of RNA and subsequent nuclear import of mature U snRNAs are also signal-mediated events. Moreover, transport of each RNA class (tRNA, rRNA, 5 S rRNA, U snRNA and mRNA) is saturable and appears to require soluble proteinaceous factors [4,5,12]. Throughout the transport pathway, RNA is bound by distinct proteins and critical signals for export are predicted to reside on these RNA-binding proteins. This has been demonstrated in studies of the HIV-1 protein Rev, which contains two discrete functional domains: a LR-NES domain and an RNA-binding domain that recognizes a structural motif in the unspliced viral RNA (see Chapter 10 in this volume by Harris and Hope). Export of the viral RNA is dependent on the function of the NES signal in the bound Rev protein. In terms of endogenous cellular RNAs, mRNA is bound by heterogenous nuclear proteins that form the hnRNP, and NES sequences are present in these hnRNP proteins (e.g. hnRNP protein A1 and Npl3p) [6]. Thus export of the mRNA may depend on recognition of signals in the bound proteins, and these proteins effectively serve as adapters for connection to the transport machinery. The same may also be true for export of rRNAs in ribosomal subunit complexes and U snRNPs; although identification of NESs in the respective bound proteins is still under investigation.

In addition to protein-based signals, physical features of RNA molecules themselves also contribute to efficient transport. For example, the export of U snRNAs requires recognition of their monomethylguanosine cap structures [4,5,13]. Studies have also suggested that the polyadenylated tail and guanosine cap structures may facilitate the export of mRNAs. The import of U snRNPs is dependent on both the cytoplasmic binding of Sm core proteins and recognition of the modified trimethylguanosine cap structure of the RNA [4,5,13]. In general, these RNA signals may ultimately serve as binding sites for adapter proteins harbouring transport signals.

Transport receptors, adapters and Ran

The identification of localization signals strongly suggested that receptors bind to these sequences and direct transport. Furthermore, the unexpected heterogeneity of transport signals indicated that an equally extensive repertoire of receptors also exists. Until recently, identification of the nucleocytoplasmic transport machinery proved difficult. Approaches that combined *in vitro* protein-import assays and fractionation of cell cytosol identified three soluble proteins that are required for cNLS-mediated nuclear import: an adapter protein (importin-α/karyopherin α/NLS receptor/Srp1/Kap60), a transport receptor (importin-β/karyopherin β1/p97/PTAC97Kap95/Rsl1) and the GTPase Ran (Gsp1p in yeast) [4,5,13–16]. Each protein performs a prototypical function in the import pathway. Subsequent studies have shown that other transport pathways utilize a family of related transport receptors, additional adapter proteins, and a variety of Ran-associated factors.

Transport receptor family

By searching for protein-sequence homology with the first reported transport receptor, importin/karyopherin β1, multiple groups described a superfamily of yeast and vertebrate proteins (Table 2) [4,13,15,16]. Completion of the yeast *Saccharomyces cerevisiae* genome-sequencing project resulted in the definitive identification of 14 proteins in the yeast family. In addition to sequence homology, most have been tested experimentally for function, and shown to

Table 2. Transport receptors

Multiple names for a single polypeptide reflect its coincident identification by several laboratories. The putative transport receptor Kap120 is based on homology; transport function and substrates have not been reported. SGD ORF, Stanford Genome Database Open Reading Frame; TS, temperature-sensitive for growth. References: [4,5,8,9,13,15–18,20–22,24] and the SGD.

Yeast family names				
Names	SDG ORF	Null strain phenotype	Vertebrate homologues	Transport cargo
Import mediators				
Kap95/Rsl1	ylr347c	Lethal	Importin β/karyopherin β1/p97/PTAC97	Importin/karyopherin α as an adapter for cNLS substrates; Snurportin as an adapter for U snRNP HIV-1 Rev and Tat, HTLV-1 Rex
Kap104	ybr017c	TS	Transportin	hnRNP proteins (yeast Nab2 and Nab4/Hrp1, vertebrate A1 and F)
Kap121/Pse1	ymr308c	Lethal	RanBP5/karyopherin β3	Pho4, ribosomal proteins
Kap123/Yrb4	yer110c	Viable	RanBP6/karyopherin β4	Ribosomal proteins
Mtr10	yor160w	TS	?	Npl3 (mRNA-binding protein)
Sxm1/Kap108	ydr395w	Viable	RanBP7, RanBP8	Lhp1
Pdr6/Kap122	ygl016w	Viable	?	TFIIA
Kap114	ygl241w	Viable	?	TATA-binding proteins
Nmd5/Kap119	yjr132w	Viable	RanBP7, RanBP8	TFIIS, Hog1
Export mediators				
Cse1	ygl238w	Lethal	CAS	Importin/karyopherin α1
Crm1/Xpo1	ygr218w	Lethal	Crm1/exportin 1	Subset of LR-NES proteins
Los1	ykl205w	Viable	Exportin-t	tRNA
Msn5/Ste21	ydr335w	Viable	?	Pho4
Putative transport receptors				
Kap120	ypl125w	?	?	?

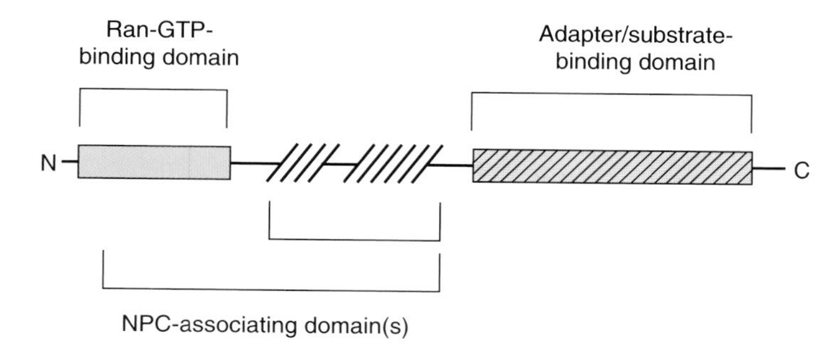

Figure 1. Schematic of a typical importin/karyopherin β1-like transport receptor [13,15–17]
Each member of the transport receptor family contains three related domains. The N-terminal region is the most homologous region amongst the members, and is where the putative Ran-GTP-binding domain is located. Two other functional domains are present, the NPC-associating and the adapter/substrate-binding domains.

facilitate nucleocytoplasmic transport of particular substrates in a signal-dependent manner (Table 2).

All receptors in this family contain three transport-related domains (Figure 1). Each receptor associates with their cognate transport substrates by binding either directly the *cis*-acting NLS or NES elements, or indirectly via adapter proteins. These substrate-associating domains (Figure 1) are not highly related in primary amino acid sequence. This finding is commensurate with these receptors interacting with diverse substrates, ranging from tRNA to proteins with distinct NLSs and NESs (Table 2). Overall, the transport of different classes of nuclear proteins by different pathways can be defined by the identity of the transport receptor involved.

Docking of the transport substrates–receptor complexes at the NPC and navigation through the portal is presumably governed by transient interactions with nucleoporins (see Chapter 7 in this volume) [17,18]. Such interactions are achieved, at least in part, by the transport receptors through their nucleoporin-associating domain(s) (Figure 1). Finally, both import and export processes utilize the GTPase Ran (see below) [14], presumably by direct binding of Ran to nucleoporins and to the transport receptors (Figure 1). Interestingly, despite the fact that there are three domains common to all transport receptors, two functionally distinct classes of receptor have emerged: those that promote nuclear import and those that promote nuclear export (Table 2).

Adapters

Adapters are defined as proteins that bridge the interaction between transport substrate and receptor. Adapters can also be formally considered transport substrates; however, they shuttle continuously between the nuclear and cytoplasmic compartments. The archetypal adapter in the protein-import pathway is importin/karyopherin α1 [3,16]. Importin/karyopherin α1 has at least three functional domains; one that interacts directly with proteins containing cNLS signals, another that interacts for import with the transport receptor importin/karyopherin β1, and an NES region [3,5,13]. The minimal 41-amino-acid region of importin/karyopherin α1 that is required for importin/karyopherin β1 binding (termed the IBB region) acts as an NLS [13,16]. Fusion of the IBB to a heterologous protein is sufficient for nuclear import. Interestingly, some importin/karyopherin β1 import substrates interact directly with the β subunit (e.g. [8,9]), and other import receptors also interact directly with the transport substrate (Table 2). Therefore, in multiple transport pathways the requirement for such adapters is bypassed.

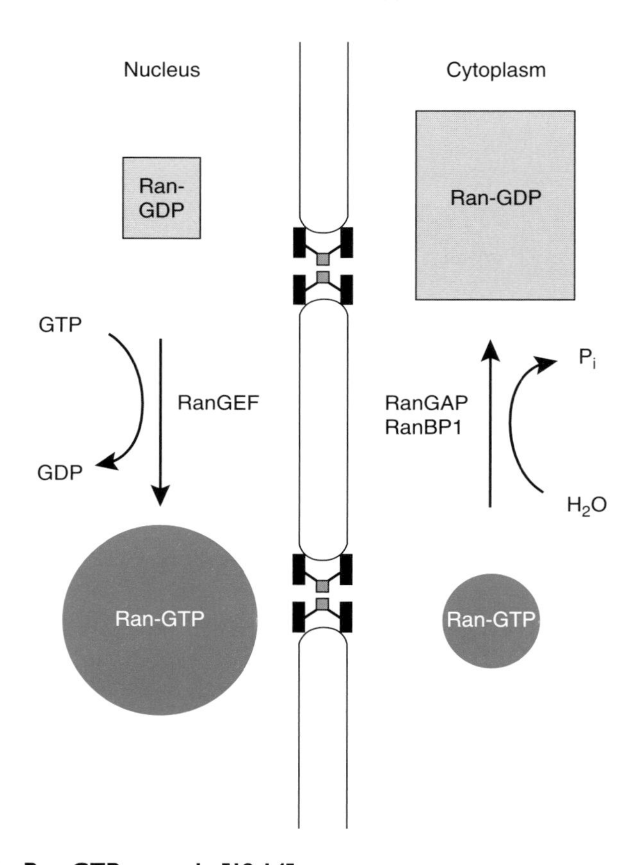

Figure 2. The Ran-GTPase cycle [12,14]
The small GTPase Ran cycles between GTP- and GDP-bound forms; however, its intrinsic GTPase activity is low. RanGAP and RanBP1 are two Ran accessory proteins that stimulate GTP hydrolysis to yield Ran-GDP, and they are localized in the cytoplasm. The Ran guanine nucleotide-exchange factor, RanGEF, binds to chromatin, and is restricted to the nucleus. Because the concentration of GTP is considerably greater than that of GDP in the cell, RanGEF primarily promotes the exchange of Ran-GDP to Ran-GTP. Based on the subcellular localization of these effector proteins, Ran is predicted to be in the GTP-bound form in the nucleus and in the GDP-bound form in the cytoplasm.

The relatively simple model of importin/karyopherin α1 acting as an adapter between cNLS substrates and importin/karyopherin β1 may be elaborated further to include multiple layers of adapters. For endogenous mRNA, the hnRNP proteins are likely candidates for adapters. However, an export receptor for the M9 NES in hnRNP A1 has not been reported. It is possible that the receptor for the M9 NES may in effect be an adapter with a distinct NES, and recognition of the adapter's NES could direct export. These additional layers of adapters may be required to facilitate NPC interactions. For example, two LR-NES-containing factors have been identified in yeast (Gle1 and Mex67) that interact with nucleoporins and are required for mRNA export [10,11].

Ran and Ran-binding proteins

Ran is a small GTPase, belonging to the Ras superfamily, that is essential for nucleocytoplasmic translocation [12,14,16]. Although intrinsic GTPase activity is low, Ran alternates between the GTP-bound and GDP-bound forms when activated by effector proteins. Binding of two cytoplasmic proteins, Ran-GTPase-activating protein (RanGAP) and RanBP1, stimulates GTP hydrolysis 100000-fold. In contrast, the nuclear protein RCC1, a Ran guanine nucleotide-exchange factor, or RanGEF, catalyses the conversion of Ran-GDP into Ran-GTP (Figure 2). The differential subcellular distribution of these critical regulators suggests strongly that cytoplasmic Ran is mainly in the GDP-bound form, whereas nucleoplasmic Ran is maintained in the GTP-bound state (Figure 2). Nuclear Ran-GTP is utilized in both the import and export pathways, and treatments that diminish the concentration of nuclear Ran-GTP inhibit nucleocytoplasmic transport [12,14]. These effects are probably mediated by Ran-GTP binding to transport receptors. Ran-GTP binding to importin/karyopherin β1 at the nucleoplasmic face of the NPC is considered one of the last steps in cNLS-mediated import (Figure 3) [4,5,13,16]. In contrast, nuclear association of Ran-GTP with several of the export receptors and their cognate substrates may promote export through the NPC (see below).

Nucleocytoplasmic transport models

Precisely how substrates, transport receptors, adapters, Ran and Ran-binding proteins interact to promote nucleocytoplasmic movement is under intense investigation. Transport in either direction proceeds by an ordered, multi-step process. First, substrates interact with transport receptors, either directly or indirectly, through adapter proteins. Secondly, to gain access to either the cytoplasm or nucleoplasm, transport complexes must dock at the NPC and then move through the portal. Finally, once transported to their destination, receptor–substrate complexes disassemble, and transport factors are recycled back to the former compartment. A combination of all the factors described in

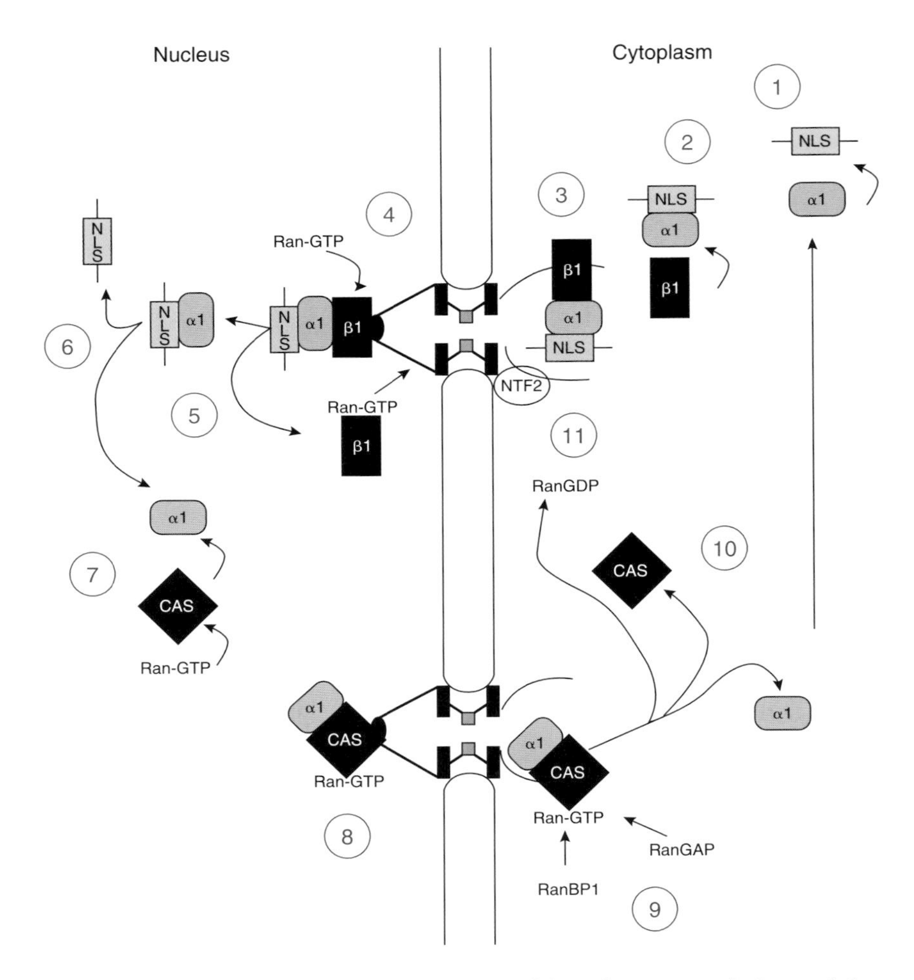

Figure 3. cNLS import and subsequent export of the adapter protein importin/karyopherin αI [4,14–16,20]

Import of a cNLS-containing protein begins with the association between the adapter protein, importin/karyopherin αI (αI), and the substrate (step 1). The transport receptor importin/karyopherin βI (βI) binds to the adapter, and docks at the NPC via interactions with specific nucleoporins (steps 2 and 3). The transport complex moves through the portal in an energy-dependent manner. Once at the nucleoplasmic face of the NPC, nuclear Ran-GTP binds to importin/karyopherin βI (step 4). This association releases the cNLS substrate–adapter dimer into the nucleoplasm, and importin/karyopherin βI exits the nucleus, presumably associated with Ran-GTP (step 5). By an unknown mechanism, importin/karyopherin αI and the cNLS-containing protein dissociate; the substrate is localized to its site of action and importin/karyopherin αI becomes a substrate for nuclear export (step 6). A distinct transport receptor, CAS in vertebrate cells, binds simultaneously to importin/karyopherin αI and Ran-GTP (step 7). The trimeric complex docks at the NPC (step 8), and moves through it. At the cytoplasmic face, the complex is most likely disassembled by the conversion of Ran-GTP into Ran-GDP (step 9; see also Figure 2). In the cytoplasm importin/karyopherin αI is free to initiate another round of import and CAS is recycled back to the nucleus (step 10). The function of Ran-GDP in the cytoplasm is not clear. Ran-GDP is targeted to the cytoplasmic fibrils of the NPC, and associates with a small protein, nuclear transport factor 2 (NTF2) (step 11). NTF2 may shuttle Ran back to the nucleus and/or help stabilize import-complex formation/translocation [16].

the previous sections are utilized during nuclear transport. Here we will consider the import of a cNLS-containing substrate, and the subsequent export of the adapter protein, importin/karyopherin α1 (Figure 3).

cNLS import and recycling of importin/karyopherin α1 through the NPC

In the cytoplasm, importin/karyopherin α1 binds directly to both the substrate cNLS and the transport receptor importin/karyopherin β1 [3,16]. This heterotrimeric complex docks at filaments on the cytoplasmic face of the NPC by direct interaction of the transport receptor with a subset of nucleoporins (see Chapter 7 in this volume). Once docked, the trimeric complex moves through the central channel of the NPC to the nuclear face. This represents a distance of up to 500 Å or more. The mechanism for how this distance is crossed is unresolved.

Macromolecules without an NLS diffuse passively through the NPC at rates inversely proportional to their mass, reflecting size restrictions of the channel [4]. In contrast, the transport receptor complexes are actively translocated without regard to size. Most models for NPC translocation have proposed some type of facilitated diffusion of the receptor complex that is mediated by association and disassociation with particular nucleoporins [19]. Based on recent studies of nucleoporin substructural localization, there are potential receptor-binding sites located throughout the pore channel [18,19]. How the transport receptor–nucleoporin binding events result in propulsion through the channel is unknown. A series of binding reactions with graded affinities along nucleoporin 'tracks' seems unlikely given the symmetrical localization of most nucleoporins and the shuttling properties of the transport receptors [16,19]. The vectorality for Ran-GDP/GTP (Figure 2) may play a key role (see below). Studies analysing the mechanism of movement are of upmost importance.

After movement through the pore channel, the nucleoplasmic NPC face serves as the release site for import complexes. Binding of nuclear Ran-GTP to importin/karyopherin β disrupts the interaction with importin/karyopherin α1 [4,5,15,16]. Both importin/karyopherin β1 and importin/karyopherin α1 are then recycled back to the cytoplasm to support additional rounds of nuclear import. It is unclear how importin/karyopherin β moves back to the cytoplasm. Interestingly, a distinct transport receptor, termed CAS, binds importin/karyopherin α1 and Ran-GTP simultaneously and mediates importin/karyopherin α export (Figure 3) [20]. Studies *in vitro* demonstrate that CAS associates with a unique set of nucleoporins and, similar to the import pathway, it is thought that the export complexes move through the NPC by association with nucleoporins. Upon reaching the cytoplasmic NPC face, the putative ternary export complex (importin/karyopherin α1–CAS–Ran-GTP) is disassembled by the conversion of Ran-GTP into Ran-GDP (Figure 3) [20]. GTP hydrolysis is stimulated by the cytoplasmic presence of RanGAP and RanBP1. The low affinity of CAS for Ran-GDP effectively prevents importin/karyopherin α1 complex formation and substrate re-import is minimized. CAS must then promote its own recycling back to the nucleus.

The shuttling of importin/karyopherin α1 illustrates the phenomenal interdependency of the import and export pathways. Perturbations that disrupt the export of importin/karyopherin α1 will have a direct negative impact on the import of cNLS-containing proteins. Because so many nuclear proteins utilize the cNLS for localization, a wide-range of cellular functions will be disturbed. Indeed, mutations in the gene encoding the *S. cerevisiae* yeast homologue of CAS, Cse1p, result in nuclear accumulation of importin/karyopherin α1 and defects in mRNA export [21]. One explanation is that importin/karyopherin α1 is not available in the cytoplasm to import essential mRNA export factors that utilize the cNLS import pathway. Alternatively, Cse1p could be involved directly in mRNA export. It is important to distinguish primary and secondary effects of disrupting particular transport pathways.

Directionality and energetics

A major question in the field of nuclear transport is how a substrate is transported preferentially in one direction. All available data suggest that transport receptors are restricted to promoting substrate translocation in one direction only. For example, transportin binds the M9 NLS in cytoplasmic hnRNP A1 to promote import. However, transportin does not appear to export the shuttling hnRNP A1, despite the finding that the same M9 sequence can be utilized for translocation in both directions [5]. Therefore, the specificity of the receptor for a given signal is a key determining event in targeting a substrate for import or export.

Recently, a number of studies have documented that localization signals, and therefore transport receptor binding, can be regulated in response to developmental or environmentally stimulated signalling pathways. This is nicely illustrated with the yeast transcription factor Pho4. Pho4 is differentially phosphorylated in response to levels of environmental phosphate, and the protein-phosphorylation state correlates with the direction of nuclear transport [22]. Specifically, under phosphate-starvation conditions, unphosphorylated Pho4 has a high binding affinity for a cytoplasmic transport receptor (Pse1/Kap121) that results in nuclear import and transcriptional activation [22]. When environmental phosphate is readily available, nuclear Pho4 is specifically phosphorylated, resulting in recognition by a distinct export receptor (Msn5) and subsequent localization to the cytoplasm. Importantly, cytoplasmic Pse1/Kap121 does not bind to phosphorylated Pho4, and therefore the substrate is not re-imported. Thus the direction of transport can be regulated.

The compartment-specific Ran-GTP/GDP gradient plays a major role in controlling directionality [14,16]. Import complexes are assembled in the cytoplasm in the presence of Ran-GDP, and are disassembled in the nucleus in response to Ran-GTP. In contrast, nuclear Ran-GTP binding stabilizes export complexes, and hydrolysis to GDP in the cytoplasm results in cargo release. Some have argued that the Ran-GTP-based vectorality is sufficient for directionality; however, not all nuclear-transport pathways require Ran [16]. The architecture of the NPC may also play a role in directionality [19]. The cytoplasmic and nucleoplasmic faces of the NPC are markedly different structurally. Peripheral filaments protrude outward from the cytoplasmic face, whereas the filaments on the nucleoplasmic face are organized into an elaborate basket-like structure. Moreover, the nucleoporin composition of the cytoplasmic filaments and nucleoplasmic basket is unique [19].

It is widely accepted that energy is required for nucleocytoplasmic transport [4,16]. However, identifying the source of this energy has proved quite challenging. Experiments demonstrating that Ran and GTP are required for nuclear import led to the suggestion that Ran-GTP hydrolysis might fuel nuclear transport [14]. However, translocation through the NPC does not require hydrolysis of any nucleotide triphosphate, and the energy derived from GTP hydrolysis by Ran is not required for movement [16]. Transport pathways that do not utilize Ran, possibly mRNA export, may have even more complex energetic requirements.

Conclusions and future directions

Although the wealth of known signals, receptors, adapters and accessory proteins required for nucleocytoplasmic translocation is impressive, many issues need to be investigated further. Pinpointing any energy source(s) required during NPC translocation, determining additional contributions of Ran and unravelling the mechanism for movement through the NPC channel are of high priority.

It is very likely that distinct transport pathways utilizing uncharacterized signals and receptors exist. At present, only one importin/karyopherin $\alpha 1$ homologue has been identified in yeast, whereas five homologous vertebrate proteins have been reported [15,16]. It is not known if the 5-fold expansion in αs will reflect the same amount of diversity of βs in vertebrate cells, or if selective vertebrate transport is achieved through a variety of specialized adapter proteins that recognize a wide range of unique NLSs and/or NESs.

Compared with protein transport, very little is known about mRNA and ribosomal-subunit nuclear export. These pathways involve transporting large hetero-oligomeric substrates that must navigate through the NPC. For example, mRNA associates with at least 20 hnRNP proteins, and several exit the nucleus with the mature transcript. Likewise, 49 large ribosomal proteins with three rRNAs and 33 small ribosomal proteins with one rRNA associate to form the exiting ribosomal subunits. These types of transport may be mechanistically distinct from that for monomeric polypeptides. Efforts during the next few years will be focused on refining further our understanding of this essential cellular process.

Summary

- *Proteins transported into and out of the nucleus require amino acid motifs called NLSs and NESs, respectively.*
- *The amino acid sequences of these signals vary considerably.*
- *A superfamily of transport receptors has been identified and each member contains three transport-related domains.*
- *Transport receptors bind to the signal sequences, either directly or through adapter proteins, to promote nucleocytoplasmic transport.*
- *The diversity of signals, receptors and adapter proteins suggests that there are many pathways for nuclear entry or exit.*
- *The direction of transport (into or out of the nucleus) is regulated in part by the small GTPase Ran as well as by intrinsic substrate motifs.*

We thank our colleagues Kathy Ryan and Albert Ho for comments on the manuscript. This work was supported by a National Research Service Award (to D.M.B.) and by grants from the National Institute of General Medical Sciences and the American Cancer Society (to S.R.W.).

References

1. Kalderon, D., Roberts, B.L., Richardson, W.D. & Smith, A.E. (1984) A short amino acid sequence able to specify nuclear location. *Cell* **39**, 499–509
2. Dingwall, C. & Laskey, R.A. (1991) Nuclear targeting sequences – a consensus? *Trends Biochem. Sci.* **16**, 478–481
3. Conti, E., Uy, M., Leighton, L. et al. (1998) Crystallographic analysis of the recognition of a nuclear localization signal by the nuclear import factor karyopherin α. *Cell* **94**, 193–204
4. Mattaj, I.W. & Englmeier, L. (1998) Nucleocytoplasmic transport: the soluble phase. *Annu. Rev. Biochem.* **67**, 265–306
5. Izaurralde, E. & Adam, S. (1998) Transport of macromolecules between the nucleus and cytoplasm. *RNA* **4**, 351–364
6. Fischer, U., Michael, W.M., Luhrmann, R. & Dreyfuss, G. (1996) Signal-mediated nuclear export pathways of proteins and RNAs. *Trends Cell Biol.* **6**, 290–293
7. Michael, W.M., Eder, P.S. & Dreyfuss, G. (1997) The K nuclear shuttling domain: a novel signal for nuclear import and nuclear export in the hnRNP K protein. *EMBO J.* **16**, 3587–3598
8. Palmeri, D. & Malim, M.H. (1999) Importin β can mediate the nuclear import of an arginine-rich nuclear localization signal in the absence of importin α. *Mol. Cell. Biol.* **19**, 1218–1225
9. Truant, R. & Cullen, B.R. (1999) The arginine-rich domains present in human immunodeficiency virus type 1 Tat and Rev function as direct importin β-dependent nuclear localization signals. *Mol. Cell. Biol.* **19**, 1210–1217
10. Murphy, R. & Wente, S. (1996) An RNA export mediator with an essential nuclear export signal. *Nature (London)* **383**, 357–360
11. Segref, A., Sharma, K., Doye, V. et al. (1997) Mex67p, a novel factor for nuclear mRNA export, binds to both poly(A)+ RNA and nuclear pores. *EMBO J.* **16**, 3256–3271

12. Dahlberg, J.E. & Lund, E. (1998) Functions of the GTPase Ran in RNA export from the nucleus. *Curr. Opin. Cell Biol.* **10**, 400–408
13. Weis, K. (1998) Importins and exportins: how to get in and out of the nucleus. *Trends Biochem. Sci.* **23**, 185–189
14. Moore, M.S. (1998) Ran and nuclear transport. *J. Biol. Chem.* **273**, 22857–22860
15. Pemberton, L.F., Blobel, G. & Rosenblum, J.S. (1998) Transport pathways through the nuclear pore complex. *Curr. Opin. Cell Biol.* **10**, 392–399
16. Görlich, D. & Kutay, U. (1999) Transport between the cell nucleus and the cytoplasm. *Annu. Rev. Cell Dev. Biol.* **15**, 607–660
17. Wozniak, R.W., Rout, M.P. & Aitchison, J.D. (1998) Karyopherins and kissing cousins. *Trends Cell Biol.* **8**, 184–188
18. Ryan, K.J. & Wente, S.R. (2000) The nuclear pore complex: a protein machine bridging the nucleus and cytoplasm. *Curr. Opin. Cell Biol.* **12**, 361–371
19. Rout, M.P., Aitchison, J.D., Suprapto, A. et al. (2000) The yeast nuclear pore complex: composition, architecture, and transport mechanism. *J. Cell Biol.* **148**, 635–651
20. Kutay, U., Bischoff, F.R., Kostka, S. et al. (1997) Export of importin a from the nucleus is mediated by a specific nuclear transport factor. *Cell* **90**, 1061–1071
21. Solsbacher, J., Maurer, P., Bischoff, F.R. & Schlenstedt, G. (1998) Cse1p is involved in export of yeast importin a from the nucleus. *Mol. Cell. Biol.* **18**, 6805–6815
22. Kaffman, A. & O'Shea, E.K. (1999) Regulation of nuclear localization: a key to a door. *Annu. Rev. Cell Dev. Biol.* **15**, 291–339
23. Schaap, P.J., Van't Riet, J., Woldringh, C.L. & Raue, H.A. (1991) Identification and functional analysis of the nuclear localization signals of ribosomal protein L25 from *Saccharomyces cerevisiae*. *J. Mol. Biol.* **221**, 225–237
24. Titov, A.A. & Blobel, G. (1999) The karyopherin Kap122p/Pdr6p imports both subunits of the transcription factor IIA into the nucleus. *J. Cell Biol.* **147**, 235–245

9

The control of gene expression by regulated nuclear transport

Eric D. Schwoebel and Mary Shannon Moore[1]

Baylor College of Medicine, Department of Molecular and Cellular Biology, One Baylor Plaza, Houston, TX 77030, U.S.A.

Introduction

There are numerous cases in the cell in which gene expression is controlled by regulated nuclear transport. Because transcription and translation take place in the nucleus and cytoplasm, respectively, the cell is able to control each process by what it makes available in each compartment. For example, a number of constitutively nuclear proteins are required to maintain nuclear structure and function and many of these must be imported into the nucleus after their translation in the cytoplasm. Conversely, RNAs of all types must be transported out of the nucleus to partake in translation, and this export appears to be dependent on proteins bound to that RNA (proteins that must first be imported). In addition, nuclear transport can also be utilized to control the timing of gene expression by regulating the transport of proteins that stimulate (or inhibit) transcription of a particular gene or set of genes. Since a protein (or protein fragment) cannot directly activate transcription on chromatin until it physically binds it, keeping it out of the nucleus keeps it away from chromatin. So, to put it simply, as long as the potential transcription factor is kept out of the nucleus it can't act as a transcription factor.

[1]*To whom correspondence should be addressed (e-mail: mmoore@bcm.tmc.edu)*

The nuclear and cytoplasmic compartments of the eukaryotic cell are separated by the nuclear envelope, and the only means of passage between these two compartments is through the nuclear pore complexes (NPCs). The NPCs allow diffusion of molecules smaller than approx. 50 kDa between the compartments, but larger molecules must be transported actively. The presence of a nuclear localization sequence (NLS) on a protein can result in transport of that protein into the nucleus. The classical or basic type of NLS is generally comprised of either a short stretch of basic amino acids or two short stretches separated by a spacer region of about 10 amino acids (bipartite NLS). While this type of NLS was the first one to be discovered, there now appear to be several different kinds of sequence (each with their own receptor) that can function as NLSs. Conversely, a nuclear export sequence (NES) causes export of proteins from the nucleus to the cytoplasm. One type is rich in hydrophobic amino acids (particularly leucine) but like NLSs, there are now known to be multiple forms of NES (see Chapter 8 in this volume by Barry and Wente). However, the mere presence of these signals is not always sufficient to result in directional transport of the protein. The signal must be exposed in the correct conformation so as to be accessible to the transport machinery. The activity of these signals can be modified (either positively or negatively) by various modifications in the protein structure. In this manner, the cell can regulate the localization of proteins by manipulating their modifications. In this chapter we will describe the effects of three different types of structural modification that are known to affect nuclear transport: ligand binding, phosphorylation and proteolysis.

Ligand-mediated nuclear import

In some cases the cellular localization of an NLS-containing protein is controlled by ligand binding by that protein, an example being the glucocorticoid receptor (GR). Glucocorticoids are steroid hormones produced during starvation conditions that have metabolic effects on several cell types. Immunofluorescence microscopy shows that in the absence of glucocorticoid (or dexamethasone, a glucocorticoid agonist) most of the GR is located in the cytoplasm, but migrates into the nucleus when dexamethasone is added (see Figure 1). This migration into the nucleus is rapid, with a $t_{1/2}$ of approx. 5 min following the addition of dexamethasone. How does the addition of ligand affect the nuclear transport of the GR receptor?

The GR receptor contains two NLSs encoded in the protein, designated NL1 and NL2 [1]. While there are rough consensus sequences for some NLS motifs, the ultimate proof of a region's activity is to attach the putative NLS to a cytoplasmic reporter protein and determine whether the presence of that putative NLS confers nuclear import on the reporter. NL1, the more N-terminal NLS similar to the bipartite NLS consensus sequence, is near the DNA-binding domain of the GR. Fusion of this part of the GR (amino acid residues

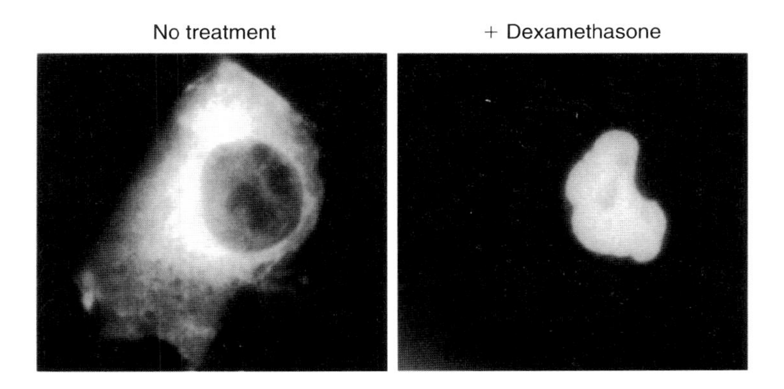

Figure 1. Cultured HeLa cells expressing green fluorescent protein (GFP)–GR incubated in the absence or presence of dexamethasone
The binding of the dexamethasone ligand to the GR stimulates transport of the receptor into the nucleus. The GFP–GR expression plasmid was donated kindly by Dr K. Yamamoto, and the experiment performed by Emily Avila.

497–524) with β-galactosidase, a protein not normally imported into the nucleus, confers nuclear import upon it regardless of the presence or absence of dexamethasone. This ligand-independent nuclear import is probably more typical of NLS-containing proteins than the ligand-dependent form. The GR is now known to contain both types.

There is a second NLS (NL2) in a region of the GR within the ligand-binding domain (amino acids 540–795) [1]. Fusion of this region of the GR with β-galactosidase confers nuclear localization of the reporter only when dexamethasone is added. The presence of these two NLSs on the protein, one constitutive and the other regulated, would presumably result in constitutive nuclear localization of the GR. Why then does GR appear cytoplasmic in cells to which ligand has not been added? It has been determined recently that the GR also contains a NES [2], which stimulates export of proteins from the nucleus. The presence of both an NES and an NLS on the same protein results in shuttling, in which the protein moves constantly between the two compartments. In the absence of ligand only the constitutive NL1 is active, and by immunofluorescence microscopy the GR appears cytoplasmic. In this case nuclear export of ligand-free GR directed by the NES is more rapid than the rate of import stimulated by NL1. The fact that the GR is shuttling can be demonstrated by the addition of leptomycin B, an inhibitor of certain types of nuclear export. In the absence of ligand, addition of leptomycin B increases the amount of GR localization in the nucleus, indicating that maintenance of the high levels of GR in the cytoplasm in the absence of ligand is dependent on nuclear export. Thus the changes in steady-state GR localization in the presence and absence of ligand are probably due to a change in the relative rates of import and export (Figure 2). In effect, while the single NES is more efficient

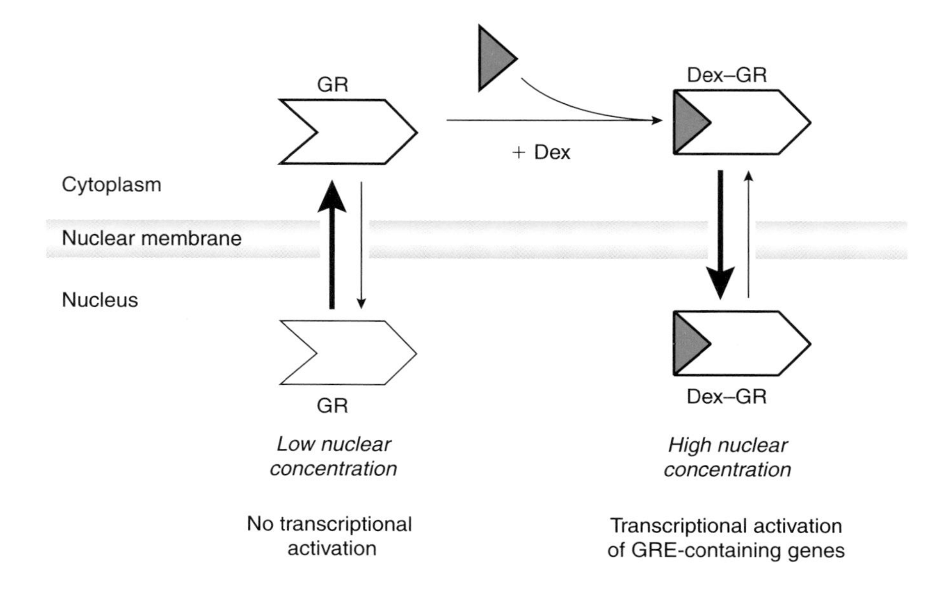

Figure 2. The GR shuttles between the nuclear and cytoplasmic compartments
In the absence of ligand, nuclear export is more efficient than nuclear import. However, in the presence of ligand (dexamethasone, Dex), the nuclear-import reaction is favoured, and the nuclear concentration of Dex–GR increases. In the nucleus, the Dex–GR is capable of binding to DNA and initiating transcription of genes containing glucocorticoid-responsive elements (GREs)

than NL1 alone, the NES is less efficient at inducing export than the combined effects of NL1 and NL2 at inducing import in the presence of ligand. Shuttling may allow low-level constitutive expression in the absence of hormone, but be able to respond quickly to a rise in hormone levels by increasing the proportion of nuclear receptor.

There are a number of ways in which ligand binding could differentially affect the rates of import versus export. Ligand binding could affect the exposure or conformation of NL2 or the NES directly. Alternatively, ligand binding to GR could affect its affinity for a number of proteins, such as import or export factors (for example, karyopherin α or CRM1, see Chapter 7 in this volume by Rout and Aitchison). Changes in these affinities could also affect the rates of nuclear transport. In any case, the net effect of the addition of ligand is a dynamic change in the cellular localization of GR, and consequently the expression of genes controlled by this transcription factor.

Effects of phosphorylation on nuclear import

Proteins can also be covalently modified by cellular enzymes, and this modification can affect their activity. One very common post-translational modification is phosphorylation. In some cases changing the phosphorylation state of a protein can also change its cellular location. SWI5 is an example of such a protein. SWI5 is a transcription factor in yeast that stimulates transcription of a family of genes including *HO*, an endonuclease involved in mating-type switching in *Saccharomyces cerevisiae*. Utilizing synchronized cells, it was determined that SWI5 is synthesized in the S, G_2 and M phases of the cell cycle, but immunofluorescence microscopy indicates that this protein remains in the cytoplasm during these stages, and is therefore transcriptionally inactive [3]. However, during late anaphase and early G_1, SWI5 translocates to the nucleus, where it activates transcription. Note that because yeast undergo what is called closed mitosis, in which the nuclear envelope does not breakdown, the cytoplasmic and nuclear compartments remain distinct throughout the cell cycle. What regulates this nuclear translocation of SWI5 at a particular stage of the cell cycle?

Fusion of amino acids 633–682 of SWI5 to β-galactosidase confers cell-cycle-specific nuclear localization identical to that seen in the wild-type protein [3]. Sequence analysis of this region identified not only a bipartite NLS, but also two potential phosphorylation sites at Ser-646 and Ser-664, with a third reasonably close by at Ser-522 [4]. This raised the possibility that phosphorylation at one or more of these sites could affect the nuclear transport of the SWI5 transcription factor. Phosphorylation of specific residues can be determined by labelling cells with ^{32}P, isolating the protein of interest and producing peptides of that protein by digesting it with a protease such as trypsin. These peptides are separated by electrophoresis, and the presence of ^{32}P on specific peptides examined. Using such analysis on cells from different stages of the cell cycle, it was determined that all three serine residues were phosphorylated during the M (when the protein stays in the cytoplasm), but not during the G_1 (when SWI5 enters the nucleus), phase. The proximity of the phosphoserines to the NLS lends credence to this hypothesis that phosphorylation of SWI5 could inhibit its nuclear localization. This idea was supported by experiments in which one or more of the serine residues were mutated to alanine, which cannot be phosphorylated. Mutation of all three serines resulted in constitutive nuclear localization of SWI5, and its localization was most sensitive to mutation of Ser-646, which is in the midst of the bipartite NLS. Thus the ability of SWI5 to be imported into the nucleus correlates well with dephosphorylation of the cytoplasmic protein (Figure 3).

There are also examples of proteins in which phosphorylation can stimulate nuclear import [5]. In some cases the mechanism of this effect has been determined. For example, phosphorylation of protein kinase C-α stimulates nuclear import by inducing conformational changes that unmask its NLS. Phosphorylation or dephosphorylation of a protein can also stimulate the formation (or disassembly) of a protein complex. For example, nuclear localization of the Dorsal protein in *Drosophila melanogaster* is stimulated when phosphorylation induces its release from the Cactus protein in the cytoplasm. There are many examples of the nuclear transport of a protein being regulated by its phosphorylation state, and depending on the protein this modification can either stimulate or inhibit its nuclear import or export. In summary, phos-

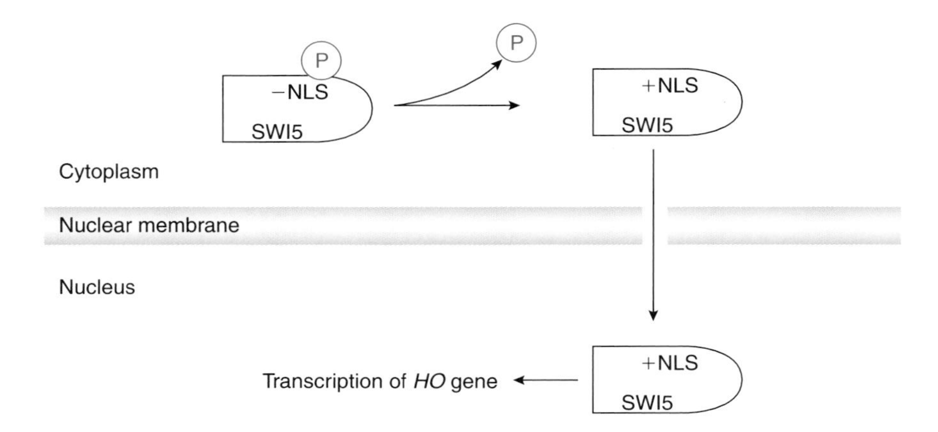

Figure 3. Nuclear import of the SWI5 transcription factor is dependent on dephosphorylation
Phosphorylation of serines in and near the NLS inactivates the NLS (−NLS) and thus inhibits nuclear localization. Dephosphorylation of these serines activates the NLS (+NLS) and stimulates nuclear localization of SWI5 and subsequent transcription of the *HO* gene.

phorylation appears to play a critical role in regulating nuclear transport in a vast number of cellular processes from T-cell activation to the establishment and maintenance of circadian rhythm.

Regulation of nuclear transport by proteolysis

Perhaps the simplest way of regulating a protein's intracellular movement is by physically tethering the protein and then releasing it in response to a signal so that it is free to move. In two different examples that we will look at, the cell accomplishes this by activating specific proteases in response to specific signals. These proteases cut off the cytoplasmic domains of selected transmembrane proteins which are then free to enter the nucleus and subsequently activate transcription.

The first example of this type of regulation was discovered in a family of proteins that play an essential role in the control of cholesterol homoeostasis in animal cells, the family of transmembrane transcription factors called sterol regulatory-element-binding proteins (SREBPs; reviewed in [6]). Investigators had known for a number of years that in response to cholesterol depletion, the rate of transcription of mRNAs coding for two key proteins involved in cholesterol metabolism was greatly increased. The amounts of both the low-density lipoprotein receptor (which binds and internalizes cholesterol from the bloodstream into the cell) and 3-hydroxy-3-methylglutaryl (HMG)-CoA reductase (the rate-limiting enzyme in cholesterol biosynthesis) were greatly increased when cholesterol dropped below a certain level in animal cells. There had to be a transcription factor (or factors) activated in response to cholesterol depletion, but it eluded purification and identification until 1993 when the first

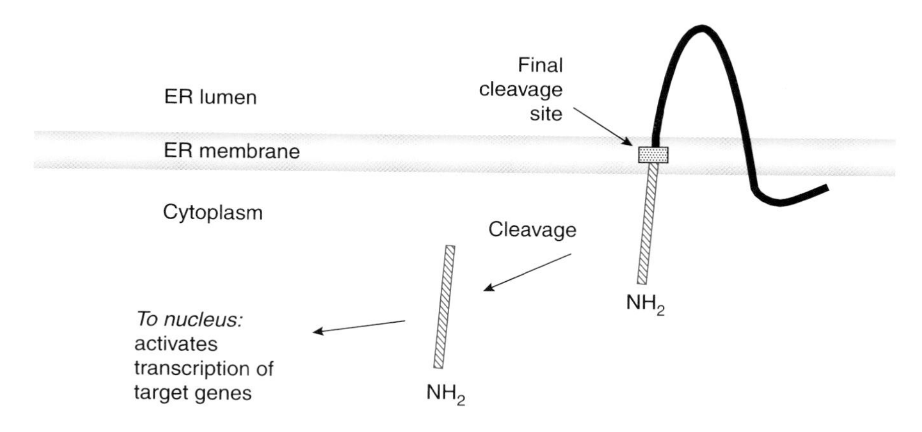

Figure 4. SREBP import requires proteolysis
Proteolysis at the cleavage site (stippled) releases the N-terminal region from the membrane-bound form of the protein. This N-terminus is then capable of nuclear entry and subsequent activation of target genes. ER, endoplasmic reticulum.

member of the SREBP family was finally identified. It was immediately apparent from its cDNA sequence and cellular location why this protein had been so difficult to purify. Investigators had been trying to isolate a soluble cellular protein only to find that the elusive SREBPs are in fact transmembrane proteins normally found anchored in membranes of the endoplasmic reticulum and nuclear envelope. These proteins are oriented such that the region of the protein that possesses transcriptional activity faces the cytoplasm. Under conditions of cholesterol (sterol) depletion, these proteins undergo a series of activated proteolytic cleavages that cut the transcriptionally active fragment away from the portion of the protein that remains membrane-bound. This fragment then enters the nucleus and turns on transcription of its target genes (Figure 4). Proof that the transcriptional activity of the SREBP family members is regulated by their proteolysis came when cells were transfected with truncated SREBP proteins lacking their transmembrane domains. These truncated SREBPs did not require proteolysis in order for them to enter the nucleus and activate transcription. As would be predicted by the model outlined here, these transfected cells were no longer responsive to their internal cholesterol levels and as a result their SREBP-responsive genes were continuously activated, resulting in an enormous build up of cholesterol within the cell.

Another example where the nuclear import of a protein is regulated by its proteolysis is found in the Notch receptor family (reviewed in [7]). This family of receptors plays numerous critical roles in the determination of cell fate during the development of a number of organisms. These receptors operate in a signalling cascade that has three parts: (i) the ligands for these receptors, the so-called DSL ligands (for Delta, Serrate, Lag-2), whose binding activates the Notch receptors, (ii) the Notch receptors themselves and (iii) the downstream effector of these receptors, the so-called CSL transcription factor [for CBF1,

Su(H), Lag-1]. What appears to happen in this signalling cascade is that, in response to binding of the DSL ligands to the Notch receptor, a protease is activated that cuts off a piece of the cytoplasmic tail of the Notch receptor. This proteolysis of a transmembrane protein is similar to what happens in the SREBP signalling pathway, except that the SREBPs are DNA-binding proteins and transcription factors in their own right, while the portion of the Notch receptor that is cleaved off is a co-activator and requires the binding of the CSL transcription factor for it to activate transcription of its target genes. What is not clear at present is where in the cell the binding of the Notch-receptor fragment and the CSL transcription factor occurs; whether the CSL protein is normally nuclear and is joined there by the cleaved Notch receptor fragment or whether these two join in the cytoplasm and translocate into the nucleus together. Also unknown is whether there are specific nuclear-transport signals that become functional upon proteolysis (in either example) or whether the nuclear translocation observed is dependent solely on the reduced size of the protein fragment. In either case, it appears that the nuclear translocation of the SREBP and Notch-receptor proteins and their subsequent transcriptional activity is regulated directly by the activated proteolysis of a membrane-bound protein.

In summary, modulation of nuclear localization provides an elegant method for regulating gene expression. In this manner it is possible to transcribe and translate a pool of transcription factor, but maintain it in a state in which it is functionally inactive by keeping it in the cytoplasm. In response to the proper signal the protein is modified, affecting its nuclear transport and ultimately its transcriptional activity. This provides a very rapid method of gene-expression control that is not dependent on intervening transcriptional/translational steps. Instances of this type of regulated import and export are being discovered daily, and it is becoming increasingly clear that regulated nuclear transport is an important method of controlling gene expression. The cell uses this method of regulating gene expression over and over again in control of the cell cycle, cell signalling and development. Only the eukaryotic cell, with its nuclear envelope separating the nucleus and the cytoplasm, is capable of nuclear transport, as the prokaryotic cell does not contain a nucleus. This ability of the eukaryotic cell to carry out nuclear transport may well explain why this cell type is capable of more complicated behaviour, a capability that was necessary for the development of multicellular organisms.

Summary

- *Many proteins show distinct nuclear- and cytoplasmic-localization patterns. For proteins above the diffusion limit of the NPC, this localization is governed by the activity of NLSs and/or NESs contained in the protein.*

- *Structural modification of proteins can affect NLS and NES activities. Ligand binding, phosphorylation and proteolysis are each capable of modifying the nucleocytoplasmic distribution of proteins.*
- *In the case of transcription factors, control of these structural modifications affects access of the transcription factor to the chromatin. This management of cellular distribution, in turn, regulates gene expression.*

Additional reading

Wente, S.R. (2000) Gatekeepers of the nucleus. *Science* **288**, 1374–1377

Kaffman, A. & O'Shea, E.K. (1999) Regulation of nuclear localization: a key to a door. *Annu. Rev. Cell Dev. Biol.* **15**, 291–339

Nakielny, S. & Dreyfuss, G. (1999) Transport of proteins and RNAs in and out of the nucleus. *Cell* **99**, 677–690

Görlich, D. & Kutay, U. (1999) Transport between the cell nucleus and the cytoplasm. *Annu. Rev. Cell Dev. Biol.* **15**, 607–660

References

1. Picard, D. & Yamamoto, K.R. (1987) Two signals mediate hormone-dependent nuclear localization of the glucocorticoid receptor. *EMBO J.* **6**, 3333–3340
2. Savory, J.G.A., Hsu, B., Loquian, I.R., Giffin, W., Reich, T., Haché, R.J.G. & Lefebvre, Y.A. (1999) Discrimination between NL1- and NL2-mediated nuclear localization of the glucocorticoid receptor. *Mol. Cell. Biol.* **19**, 1025–1037
3. Nasmyth, K., Adolf, G., Lydall, D. & Seddon, A. (1990) The identification of a second cell cycle control on the HO promoter in yeast: cell cycle regulation of SWI5 nuclear entry. *Cell* **62**, 631–647
4. Moll, T., Tebb, G., Surana, U., Robitsch, H. & Nasmyth, K. (1991) The role of phosphorylation and the CDC28 protein kinase in cell cycle-regulated nuclear import of the *S. cerevisiae* transcription factor SWI5. *Cell* **66**, 743–758
5. Jans, D.A. & Hübner, S. (1996) Regulation of protein transport to the nucleus: central role of phosphorylation. *Physiol. Rev.* **76**, 651–685
6. Brown, M.S. & Goldstein, J.L. (1997) The SREBP pathway: regulation of cholesterol metabolism by proteolysis of a membrane-bound transcription factor. *Cell* **89**, 331–340
7. Kimble, J., Henderson, S. & Crittenden, S. (1998) Notch/LIN 12 signaling: transduction by regulated protein slicing. *Trends Biochem. Sci.* **23**, 353–357

10

RNA export: insights from viral models

Matthew E. Harris and Thomas J. Hope[1]

Infectious Disease Laboratory, The Salk Institute for Biological Studies, La Jolla, CA 92037, U.S.A.

Introduction

A defining feature of eukaryotic cells is the division of the cell into a nucleus and cytoplasm. This division separates the site of RNA transcription from the site of translation, and this requires mRNAs to be exported from the nucleus via the nuclear pore to direct the synthesis of proteins in the cytoplasm.

Importantly, naked RNA is not exported. In the nucleus, RNA is associated with a number of proteins forming a ribonucleoprotein (RNP) complex. RNPs exit the nucleus by going through the nuclear pore complex, which is also composed of a large number of proteins and functions as a specific gate. Only designated proteins or RNPs are allowed to pass through the nuclear pore. Identification of the proteins involved in export – both those that bind or are associated with the RNA and those in the nuclear pore complex – is ongoing.

Experiments in *Xenopus* oocytes reveal multiple, energy-dependent export pathways for different classes of RNA [1]. In these experiments, radiolabelled RNAs are microinjected into the nuclei of *Xenopus* oocytes. After incubation, nuclei are dissected manually from the cytoplasm and RNA is extracted from both compartments. Rates of export can be determined by the relative amounts of RNA in each compartment over time. Injection of an excess of unlabelled RNA will inhibit export, presumably by saturating a limiting factor. For example, if an excess of mRNA is used, then radiolabelled mRNA will

[1]*To whom correspondence should be addressed (email: hope@salk.edu).*

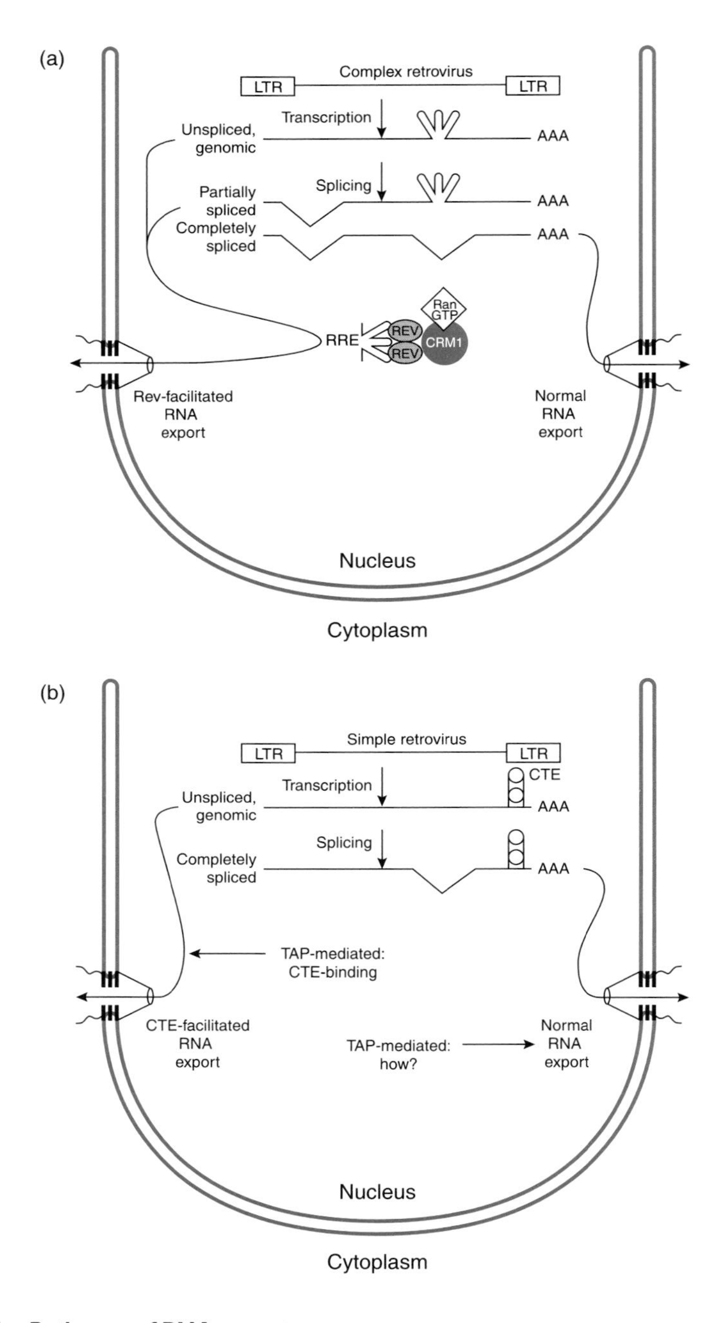

Figure 1. Pathways of RNA export

(a) Facilitated RNA export by the Rev protein of complex retroviruses. The integrated retrovirus genome, flanked by long terminal repeats (LTR), is transcribed to make an unspliced genomic RNA. This RNA contains two introns. The genomic RNA can be spliced to remove one or both introns. The completely spliced RNA is exported via the normal mRNA-export pathway. Intron-containing transcripts must be exported by a Rev-facilitated pathway. The intron-containing

(contd.)☞

not be exported but other radiolabelled RNAs continue to be exported. These competition experiments reveal multiple RNA-export pathways for mRNA, tRNA, small nuclear RNA (snRNA) and rRNA.

Additionally, it should be noted that a lack of export does not necessarily mean that export itself is defective. RNAs undergo co-transcriptional and post-transcriptional processing in the nucleus. Some of these events are prerequisites for RNA export. Thus a defect in a processing step might appear as an export defect.

Studies of the transport of unconventional viral RNA have shed light on the complicated process of RNA export. The first unconventional viral RNAs studied were those that were exported with introns. In normal RNA export, introns are removed by splicing before export can take place. In contrast, retroviruses export RNA containing one or more introns. One class, the complex retroviruses, do so using a protein called Rev (Figure 1a). Key experiments were possible with this system because Rev was virally encoded. Stimulated by the success of these studies, another class of retroviruses, the simple retroviruses, which also transport an intron-containing RNA into the cytoplasm but do not encode a viral protein to mediate this process, began to receive attention. RNA elements important for the cytoplasmic localization of the unconventional RNA were discovered. Recently, a protein that binds one of these elements and facilitates export has been identified.

Finally, another kind of viral RNA is being studied as a potential new model for RNA export. Herpes simplex virus (HSV) and hepatitis B virus (HBV) produce intronless RNAs. *Cis*-acting RNA elements have been identified in these transcripts that affect the cytoplasmic accumulation of several different transcripts. These elements are believed to act at the level of RNA export.

This chapter is divided into three major sections. In the first section, we discuss viral RNA export mediated by Rev or proteins with a related domain. The second section focuses on export of intron-containing RNA that uses cellular proteins that are not related to Rev. The third and last section is devoted to *cis*-acting RNA elements involved in the export of intronless viral transcripts.

Figure 1. Pathways of RNA export (contd.)
transcripts contain the Rev-responsive element (RRE) within the second intron. Multiple Rev molecules bind the RRE. A complex between Rev, RRE, CRM1 and Ran results in the export of the bound RNA. In the nucleus, Ran binds GTP. This form of Ran is required for export. (b) Facilitated export of intron-containing RNA of a simple retrovirus. Like the complex retroviruses, the Mason–Pfizer monkey virus genome is transcribed to make an unspliced genomic RNA. In contrast, this RNA only contains a single intron. The genomic RNA can splice to generate a product that is exported via the normal mRNA-export pathway. The constitutive transport element (CTE) is located near the 3′ end of the viral transcripts. The CTE facilitates the export of the intron-containing RNA. This facilitated pathway uses the cellular protein TAP, which binds the CTE, rather than a virally encoded protein. TAP is also involved in normal mRNA export, but its precise role has not been determined.

RNA export mediated by Rev or Rev-like proteins

The role of the Rev proteins in export was first suggested by genetic experiments with human immunodeficiency virus (HIV). HIV encodes its structural, enzymic, regulatory and accessory proteins from a compact genome of less than 10 kb. The virus uses post-translational cleavage, overlapping reading frames, alternative splicing and intronic messages to express all of these proteins. The structural and enzymic proteins are expressed from the full-length genomic transcript that contains two introns. The envelope protein is expressed from a partially spliced, single intron message. Translation of these transcripts and packaging of the viral RNA genome in the cytoplasm requires that these intron-containing RNAs bypass the normal requirement for splicing before export. In a Rev-minus virus, the intron-containing RNAs are not observed in the cytoplasm, while completely spliced messages, including the Rev transcript, accumulate in the cytoplasm. Rev binds to a specific RNA element, the Rev-responsive element (RRE), located in the second HIV intron. It was hypothesized that once bound to the RRE, Rev facilitated the export of the intron-containing RNA (Figure 1).

At steady state, Rev is observed in the nucleus. A nuclear localization signal, a short peptide motif in the protein, confers this localization. However, Rev was shown to move continuously between nucleus and cytoplasm in the following way. Human cells expressing Rev were fused to non-expressing mouse cells to form multi-nucleated cells or heterokaryons. Before fusion, Rev was localized in the nuclei of the human cells. After fusion, Rev was found in both the human and mouse nuclei. This assay demonstrated that Rev was a shuttling protein [2]. By moving between nucleus and cytoplasm, Rev could redistribute into both human and mouse nuclei in the heterokaryon cell (Figure 2). Such experiments illustrate the disadvantage of looking at protein localization as a static picture. Rev's nuclear localization reflects its equilibrium distribution but belies the fact that it shuttles. Importantly, Rev shuttles in the absence of its RNA substrate [2], lending credence to the idea that Rev is involved in exporting RNA.

To test Rev's role in RNA export, Rev and an RNA substrate were microinjected into oocyte nuclei. The substrate RNA consisted of an intron containing the RRE flanked by two exons. Splicing of the RNA occurred in the nucleus, but remarkably Rev was able to export the excised intron, known as a lariat because of its structure, following excision. The lariat, normally a highly unstable nuclear RNA, is more stable in oocyte nuclei. Export of the lariat by Rev demonstrated that Rev could act after splicing [3]. This dispelled the idea that Rev functioned by masking the presence of the intron within the RNA. Moreover, these experiments were direct evidence that Rev exported RNA.

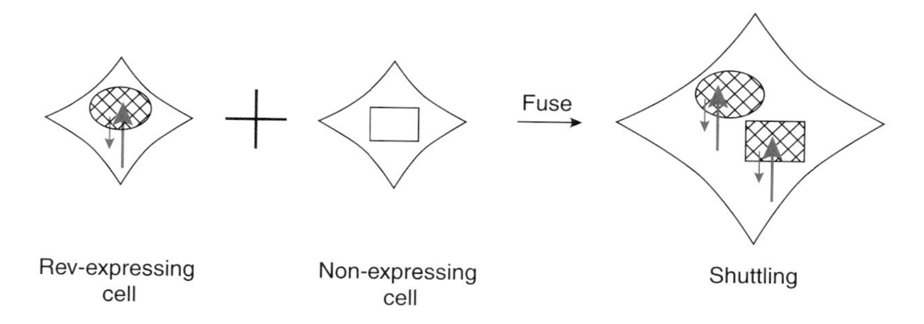

Figure 2. Rev shuttles between the nucleus and cytoplasm
The figure depicts the results of a heterokaryon assay for Rev. Rev is localized to the nucleus (hatched area) in a human cell (oval nucleus). Arrow thickness indicates the relative rates of Rev import and export. Import is faster than export, creating the impression at steady state that Rev is only in the nucleus. When this cell is fused to a mouse nucleus (square nucleus), Rev redistributes between both nuclei. This demonstrates that Rev leaves the human nucleus and imports into both nuclei. This dynamic movement is called shuttling.

Rev contains a nuclear export sequence (NES)

The N-terminus of Rev contains the nuclear localization signal, RNA-binding domain and a multimerization domain. The C-terminus of Rev contains a leucine-rich sequence that is required for function. The leucine-rich sequence was demonstrated to be a NES by microinjection studies [4,5]. A fusion protein consisting of glutathione S-transferase (GST) and the leucine-rich peptide was expressed in bacteria, purified and injected into the nuclei of somatic cells [5]. The GST-NES protein, but not GST, was rapidly exported to the cytoplasm. Similarly, the NES peptide conjugated to BSA, but not BSA alone, was exported when injected into the nuclei of *Xenopus* oocytes or somatic cells [4]. In both systems, conjugated peptides with mutations in the leucine-rich sequence remained in the nucleus. This demonstrated for the first time that the leucine-rich sequence conferred nuclear-export properties to Rev.

Along with the discovery of a NES in Rev, Wen et al. [5] showed that the cellular protein kinase A inhibitor (PKI) had an NES. A growing number of cellular proteins with Rev-like NESs are being identified; we list only a few examples in Table 1. Other large hydrophobic residues as well as leucines characterize these NESs (known or potential key hydrophobic residues in the NESs are shown in bold). Note that these hydrophobic residues adopt a similar but not identical spacing pattern in all the NESs. NES-bearing proteins are not limited to RNA export; they include cell-cycle regulators, transcription factors and structural proteins. It is thought that by mimicking a signal found in cellular proteins, Rev can interact with cellular machinery responsible for export. The discovery of the Rev NES made it possible to study and define that export machinery.

Table 1. Structure of nuclear export signals

Abbreviations used: EBV, Epstein-Barr virus; HTLV, human T-cell leukaemia virus

Viral NESs	
HIV-1 Rev	**L**QLPP**L**ER**L**T**L**D
HTLV-1 Rex	**L**SAQ**L**YSS**L**S**L**D
HSV ICP27	**I**DM**LI**D**L**G**L**D**L**S
EBV Mta	T**L**PSP**L**AS**L**T**L**E
Influenza A NS2	**ILL**R**M**SK**M**Q**L**E
Influenza B NS2	**I**EWR**M**KK**M**A**I**G
Cellular NESs	
PKI	**L**A**L**K**L**AG**L**D**I**N
p53	**M**FRE**L**NEA**L**E**L**K
Map kinase kinase	**L**QKK**L**EE**L**E**L**D
c-abl	**L**ESN**L**RE**L**Q**I**C
cyclin B1	**L**CQA**F**SD**VIL**A
α-Actin (NES1)	**L**PHA**I**MR**L**D**L**A
α-Actin (NES2)	**I**KEK**L**CY**V**A**L**D

Rev accesses a specific RNA-export pathway in the cell [4]

RNA export is a saturable process [1]. Microinjection of oocyte nuclei with an excess of unlabelled NES-conjugated BSA inhibited the export of labelled conjugates. Export of different RNAs was then tested under these inhibiting conditions. Excess NES-conjugated BSA inhibited the export of U snRNAs and 5 S rRNAs but had no effect on mRNA or tRNA export. Thus, Rev-mediated export uses cellular machinery that exports U snRNAs and 5 S rRNA. The transcription-factor protein TFIIIA, known to be involved in 5 S rRNA export, was shown subsequently to contain a Rev-like NES.

A significant advance in the RNA-export field was the identification of the Rev nuclear-export-signal receptor [6,7]. It was hypothesized that for Rev to carry out its function, a cellular protein must recognize and interact with the NES. Leptomycin B (LMB), an anti-microbial drug, was the key to identifying the NES receptor. Treatment of cells with LMB specifically blocks the export of Rev-dependent HIV RNAs but not the Rev-independent transcripts [8]. LMB also inhibits the export of Rev and U snRNAs in oocytes [6]. Furthermore, LMB inhibits the growth of the baker's yeast *Saccharomyces cerevisiae* but not the fission yeast *Schizosaccharomyces pombe*. LMB resistance in *S. pombe* maps to a gene called *CRM1* or chromosome maintenance region 1. *In vitro*, Rev binds CRM1, and this binding is dependent on a functional NES in Rev; LMB disrupts this interaction [6,7]. These data suggest that CRM1 is the NES receptor. If CRM1 is the limiting factor in oocyte nuclei, then excess CRM1 should stimulate Rev and U snRNA export and reverse the inhibition of LMB. These predictions were found to be true and defined

CRM1 as a key factor in NES-dependent export [6]. CRM1 is also known as exportin 1 to emphasize its role in export.

CRM1 is a member of the β-importin superfamily. Members of this superfamily have in common a domain that binds the small GTPase Ran. Ran is a key factor in the bi-directional transport of macromolecules between the cytoplasm and nucleus. Import and export complexes, including the CRM1–Rev complex, interact with Ran (Figure 1). Binding of GTP or GDP to Ran sends signals to transport complexes revealing whether they are in the nucleus or cytoplasm.

Other viruses encode Rev-like proteins. For example, human T-cell leukaemia virus encodes a NES-containing protein called Rex that binds to an element in the viral long terminal repeats. ICP27 from HSV, Mta from Epstein–Barr virus and non-structural protein 2 (NS2) or NEP from influenza virus have all been reported to encode NES-containing proteins [9–11] (Table 1). ICP27 appears to export intronless RNA [10,11]. Interestingly, CRM1 does not appear to be involved in cellular mRNA export.

Rev-independent facilitated export

Simple retroviruses, like the complex retroviruses, must bypass the splicing requirement for export, but there is one important difference (Figure 1b). These viruses do not encode a viral protein involved in RNA export. Instead, their RNAs contain binding sites for cellular export proteins. Studies of these viruses have focused on the identification of a *cis*-acting RNA element required for intron-containing export in the type-D retroviruses. An element was found in the Mason–Pfizer monkey virus (MPMV) and the related simian retroviruses types 1 and 2. This element, known as the constitutive transport element (CTE), has a key role in the export of cellular RNA-export proteins. Similar elements in Rous sarcoma virus, an avian retrovirus, and murine leukaemia virus that mediate cytoplasmic accumulation of the genomic RNAs have been identified [12,13].

The CTE is a small RNA sequence located at the 3′ end of MPMV [14]

Using mutagenesis and nucleic acid-modifying reagents, a secondary structure in the CTE was identified [15]. The CTE consists of two stem loops, one on top of the other. The loops have the same primary sequence but are rotated 180° relative to each other and are required for function (Figure 3). The loops were proposed to be the binding site for a cellular factor that mediates function. While the CTE is present in both the unspliced and spliced viral RNAs (Figure 3), it is only required for export of the unspliced transcript [16].

The evidence that the CTE was a genuine RNA-export element again came from microinjection of *Xenopus* oocytes [17]. The CTE was placed in the intron of the same construct used to test Rev and the RRE. RNA transcribed from this construct was microinjected into oocyte nuclei. Like Rev, the CTE

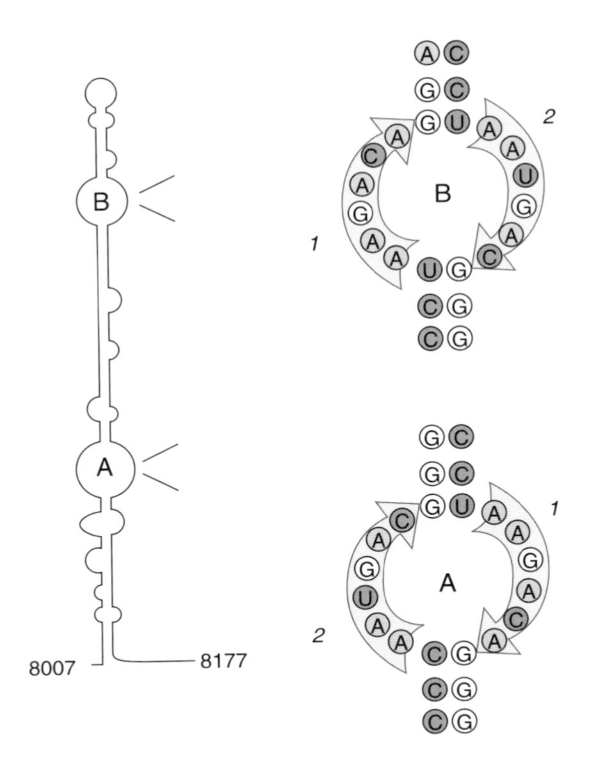

Figure 3. Structure of the MPMV CTE
A schematic of the secondary structure of the 170-bp CTE is shown on the left. The CTE contains two large loop regions of identical sequence rotated 180° relative to each other (shown by numbered arrows). The cellular protein TAP binds to the A and B loop regions.

caused export of the lariat. Therefore the CTE was then tested for its ability to inhibit other RNA export.

The CTE uses a factor involved in mRNA export

When excess CTE RNA was microinjected into nuclei, it competed with the export of radiolabelled mRNA [17]. Export of radiolabelled U snRNA, tRNA and Rev-dependent RNA was not inhibited, nor was CTE export inhibited by an excess of NES-conjugated BSA. These results are consistent with the CTE using the mRNA pathway. Furthermore, CTE export is not sensitive to LMB [18]. Therefore the CTE and Rev both export intron-containing viral RNA, but they do so via two different export pathways.

The protein TAP binds the CTE and mediates CTE-dependent RNA export

Using a radiolabelled CTE RNA, a protein from nuclear extract that bound to the CTE was identified [19]. This protein is TAP, the human homologue of the yeast protein Mex67p. Mex67p is implicated in mRNA export in yeast. TAP binds specifically to the CTE in that it does not interact with CTE loop mutants that are deficient for export of the lariat structure from oocyte nuclei. As discussed above, excess CTE RNA can titrate a limiting factor for export of mRNA from oocytes. However, mRNA export could be restored by microinjection of TAP. This demonstrates that TAP is probably the saturable factor for CTE and mRNA export in oocytes (Figure 3). RNA helicase A is also reported to bind the CTE [20]. Its role, if any, in export of MPMV RNA is not known.

Export of intronless viral RNA

Several viruses including HBV and HSV produce intronless transcripts. Not only do introns serve as retention signals until splicing is complete, but introns also have positive effects on RNA processing. When introns are inserted into cDNAs (a gene with the introns removed) and the transcript is transfected into cells, the resulting protein expression is often increased. Analysis of some of these introns reveals that they stimulate 3′-end formation of messages. 3′-End formation, which consists of cleavage of the RNA transcript followed by the addition of a polyadenosine tail, indirectly affects export because it is a prerequisite for transport. For example, the β-globin transcript requires an intron for 3′-end formation and subsequent transport to the cytoplasm. The process of splicing may also play a role in assembling factors on to the RNA for export or for other RNA-processing events. Thus intronless messages may compensate for the lack of an intron with specialized RNA elements.

HSV-1 is a nuclear-replicating DNA virus. Of more than 70 HSV-1 genes, only four of those expressed during lytic infection contain introns. Early studies of one intronless transcript, the thymidine kinase (TK) message, revealed two interesting observations. First, the TK transcript could overcome the β-globin transcript's requirement for an intron. Secondly, intron-containing chimaeric transcripts could be found in the cytoplasm with TK.

An RNA element was found in the intronless TK transcript. Liu and Mertz [21] mapped a 119-nucleotide element within the TK transcript that could rescue expression of a β-globin transcript lacking introns. This element, the pre-mRNA processing enhancer (PPE) was found to bind heterogenous nuclear (hn) RNP L [21]. Mutations of the PPE were used to establish a correlation between binding and cytoplasmic accumulation of β-globin.

One proposed mechanism for the PPE of TK was stimulation of 3′-end processing [21]. The ability of the PPE to replace an intron in β-globin argues that it has the same function as the intron to ensure efficient 3′-end formation of β-globin (see above). The observation that TK could cause intron-containing RNA to accumulate in the cytoplasm suggested an export mechanism. Otero and Hope [22] found that the TK transcript can replace Rev and the RRE in the export of unspliced HIV RNA to the cytoplasm. Could the TK transcript be mimicking the RRE by binding a Rev-like cellular factor? Otero and Hope demonstrated further that, in this assay, TK was insensitive to LMB [22]. This indicated that the TK transcript did utilize the Rev-export pathway. Nevertheless, the TK transcript could contain an RNA-export element. The relationship between the CTE and the TK-transcript export pathways remains to be determined. Recently, Huang et al. [23] found the PPE increases 3′-end formation and allows for the transport of a β-globin construct lacking an intron. This finding supports both proposed mechanisms for the PPE. Liu and Mertz [21] suggested and Otero and Hope [22] demonstrated that there are additional *cis*-acting elements in the TK transcript besides the PPE that are involved in RNA export to the cytoplasm. This finding raises the possibility that multiple functions are carried out by distinct elements within the intronless TK transcript.

A functionally similar element, known as the post-transcriptional regulatory element (PRE), has also been identified in the HBV [24]. This element, like the HSV TK coding sequence, consists of multiple *cis*-acting sequences, can facilitate the export of an intron containing RNA and stimulate 3′-end formation, and functions in an LMB-insensitive manner [18,23]. Interestingly, the PRE in the closely related woodchuck hepatitis virus (WPRE) has the ability to post-transcriptionally stimulate the expression of heterologous cDNAs up to 10-fold [25]. This ability is unique among all the viral elements discussed in this chapter.

Relevance and future goals

Studies of these elements have focused on how these viral RNAs bypass normal cellular requirements. In studying this, we illuminate normal cellular processes and generate ideas for practical applications of gene expression. Experiments with Rev and the CTE have helped define RNA-export pathways and cellular proteins involved in export. In addition the intronless RNA elements are capable of increasing gene expression, which should be of general interest to researchers.

The post-transcriptionally acting viral RNA elements that we have discussed are summarized in Table 2, together with others not mentioned here. Questions about mRNA export remain. What role does TAP play in mRNA export? Are there subsets of mRNA that use different factors for export? For intronless transcripts, the mechanisms by which the RNA elements act remain to be determined. Studying these transcripts begs the question of the positive functions of introns. For example, do introns assemble export factors that can act after splicing? Identifying the proteins that bind RNA elements from intronless transcripts may shed light on this question. Further, these elements may teach us more about RNA processing.

Table 2. Export of unconventional viral RNAs

(1) The post-transcriptional elements in TK function without ICP27 and are not sensitive to LMB. However, ICP27 should be sensitive to LMB because it contains a Rev-like NES.
(2) NS2 has not been shown to bind RNA directly. The viral protein M1 probably bridges NS2 and RNA, but NS2 may interact with the RNA via other proteins.
(3) These proteins should be sensitive to LMB because they contain a Rev-like NES, but the experiments have not been done.
(4) Five Epstein-Barr virus replication genes have been shown to require Mta for RNA export; other viral genes may also be dependent on Mta. Mapping of a specific element, if it exists, is in progress.

Virus	Unconventional RNA	RNA element	RNA binding	Is export sensitive to LMB?
Complex retroviruses: HIV	Intron-containing	Rev responsive element	Rev (viral; has a NES)	Yes
Type D retroviruses: MPMV	Intron-containing	Constitutive transport element	TAP (Cellular)	No
Mammalian HBV	Intronless	Post-transcriptional regulatory element	Unknown	No
HSV	Intronless	Pre-mRNA processing enhancer and additional elements in TK	ICP27 (viral; has a NES) and hnRNP L (cellular)	No (1)
Influenza virus	Non-polyadenylated, non-messenger, genomic RNA	See (2)	NS2 [viral; has a NES; see (2)]	Yes (3)
Epstein-Barr virus	Intronless	See (4)	Mta (viral; has an NES)	Yes (3)

Summary

- *The retroviruses export intron-containing RNA.*
- *The complex retroviruses encode a Rev protein that uses a leucine-rich NES to interact with CRM1 and the U snRNA-export pathway.*
- *Other viruses encode proteins with a Rev-like NES.*
- *The type-D retroviruses contain a CTE that binds the cellular protein TAP to export intron-containing RNA through the mRNA pathway.*
- *Intronless viral transcripts contain post-transcriptionally acting RNA elements that may compensate for the lack of an intron.*
- *The functions of elements in intronless RNA are not fully understood but may be in export and/or 3′-end processing.*

We thank Allison Bocksruker and Lynn Artale for help in preparing the illustrations. We apologise to those authors whose work was not cited owing to space limitations.

References

1. Jarmolowski, A., Boelens, W.C., Izaurralde, E. & Mattaj, I.W. (1994) Nuclear export of different classes of RNA is mediated by specific factors. *J. Cell Biol.* **124**, 627–635
2. Meyer, B.E. and Malim, M.H. (1994) The HIV-1 Rev *trans*-activator shuttles between the nucleus and the cytoplasm. *Genes Dev.* **8**, 1538–1547
3. Fischer, U., Sylvie, M., Michael, T., Corinne, H., Reinhard, L. & Guy, R. (1994) Evidence that HIV-1 Rev directly promotes the nuclear export of unspliced RNA. *EMBO J.* **13**, 4106–4112
4. Fischer, U., Huber, J., Boelens, W.C., Mattaj, I.W. & Luhrmann, R. (1995) The HIV-1 Rev activation domain is a nuclear export signal that accesses an export pathway used by specific cellular RNAs. *Cell* **82**, 475–483
5. Wen, W., Meinkoth, J.L., Tsien, R.Y. & Taylor, S.S. (1995) Identification of a signal for rapid export of proteins from the nucleus. *Cell* **82**, 463–473
6. Fornerod, M., Ohno, M., Yoshida, M. & Mattaj, I.W. (1997) CRM1 is an export receptor for leucine-rich nuclear export signals. *Cell* **90**, 1051–1060
7. Ossareh, N.B., Bachelerie, F. & Dargemont, C. (1997) Evidence for a role of CRM1 in signal-mediated nuclear protein export. *Science* **278**, 141–144
8. Wolff, B., Sanglier, J.J. & Wang, Y. (1997) Leptomycin B is an inhibitor of nuclear export: inhibition of nucleo-cytoplasmic translocation of the human immunodeficiency virus type 1 (HIV-1) Rev protein and Rev-dependent mRNA. *Chem. Biol.* **4**, 139–147
9. O'Neill, R.E., Talon, J. & Palese, P. (1998) The influenza virus NEP (NS2 protein) mediates the nuclear export of viral ribonucleoproteins. *EMBO J.* **17**, 288–296
10. Sandri, G.R. (1998) ICP27 mediates HSV RNA export by shuttling through a leucine-rich nuclear export signal and binding viral intronless RNAs through an RGG motif. *Genes Dev.* **12**, 868–879
11. Semmes, O.J., Chen, L., Sarisky, R.T., Gao, Z., Zhong, L. & Hayward, S.D. (1998) Mta has properties of an RNA export protein and increases cytoplasmic accumulation of Epstein-Barr virus replication gene mRNA. *J. Virol.* **72**, 9526–9534
12. Ogert, R.A., Lee, L.H. & Beemon, K.L. (1996) Avian retroviral RNA element promotes unspliced RNA accumulation in the cytoplasm. *J. Virol.* **70**, 3834–3843
13. King, J.A., Bridger, J.M., Gounari, F., Lichter, P., Schulz, T.F., Schirrmacher, V. & Khazaie, K. (1998) The extended packaging sequence of MoMLV contains a constitutive mRNA nuclear export function. *FEBS Lett.* **434**, 367–371

14. Bray, M., Prasad, S., Dubay, J.W., Hunter, E., Jeang, K.-T., Rekosh, D. & Hammarskjold, M.-L. (1994) A small element from mason-pfizer monkey virus genome makes human immunodeficiency virus type 1 expression and replication Rev-independent. *Proc. Natl. Acad. Sci. U.S.A.* **91**, 1256–1260
15. Ernst, R.K., Bray, M., Rekosh, D. & Hammarskjold, M.L. (1997) Secondary structure and mutational analysis of the Mason-Pfizer monkey virus RNA constitutive transport element. *Rna* **3**, 210–222
16. Ernst, R.K., Bray, M., Rekosh, D. & Hammarskjold, M.L. (1997) A structured retroviral RNA element that mediates nucleocytoplasmic export of intron-containing RNA. *Mol. Cell. Biol.* **17**, 135–144
17. Saavedra, C., Felber, B. & Izaurralde, E. (1997) The simian retrovirus-1 CTE unlike the HIV-1 RRE, utilises factors required for the export of cellular mRNAs. *Curr. Biol.* **7**, 619–628
18. Otero, G.C., Harris, M.E., Donello, J.E. & Hope, T.J. (1998) Leptomycin B inhibits equine infectious anemia virus Rev and feline immunodeficiency virus rev function but not the function of the hepatitis B virus posttranscriptional regulatory element. *J. Virol.* **72**, 7593–7597
19. Gruter, P., Tabernero, C., von Kobbe, C., Schmitt, C., Saavedra, C., Bachi, A., Wilm, M., Felber, B.K. & Izaurralde, E. (1998) TAP, the human homolog of Mex67p, mediates CTE-dependent RNA export from the nucleus. *Mol. Cell* **1**, 649–659
20. Tang, H., Gaietta, G.M., Fischer, W.H., Ellisman, M.H. & Wong, S.F. (1997) A cellular cofactor for the constitutive transport element of type D retrovirus. *Science* **276**, 1412–1415
21. Liu, X. and Mertz, J.E. (1995) HnRNP L binds a *cis*-acting RNA sequence element that enables intron-independent gene expression. *Genes Dev.* **9**, 1766–1780
22. Otero, G.C. and Hope, T.J. (1998) Splicing-independent expression of the herpes simplex virus type 1 thymidine kinase gene is mediated by three *cis*-acting RNA subelements. *J. Virol.* **72**, 9889–9896
23. Huang, Y., Wimler, K.M. & Carmichael, G.G. (1999) Intronless mRNA transport elements may affect multiple steps of pre-mRNA processing. *EMBO J.* **18**, 1642–1652
24. Huang, J. and Liang, T.J. (1993) A novel hepatitis B virus (HBV) genetic element with Rev response element-like properties that is essential for expression of HBV gene products. *Mol. Cell. Biol.* **13**, 7476–7486
25. Zufferey, R., Donello, J.E., Trono, D. & Hope, T.J. (1999) Woodchuck hepatitis virus post-transcriptional regulatory element (WPRE) enhances expression of transgenes delivered by retroviral vectors. *J. Virol.* **23**, 2886–2892

Subject index

V

X

Y